ESCAPE THE MEATRIX

ESCAPE
THE MEATRIX

EAT PLANTS, FEEL GREAT, AND SAVE THE PLANET!

STUART WALDNER

I dedicate this book to my readers, those rare and remarkable rebels willing to take the red pill and make personal changes to become healthier and create a better world.

HOUNDSTOOTH
PRESS

ESCAPE THE MEATRIX
Eat Plants, Feel Great, and Save the Planet!

ISBN   HARDCOVER: 978-1-5445-2874-8
         PAPERBACK: 978-1-5445-2875-5
         EBOOK: 978-1-5445-2876-2

AUTHOR'S NOTE

Escape the Meatrix and StuartWaldner.com are not affiliated or associated with, or authorized or endorsed by, or in any way officially connected with GRACE Communications Foundation Inc. (foodprint. org), owner of copyrights to THE MEATRIX short films, registered trademark THE MEATRIX, and domain name themeatrix.org.

DISCLAIMER: The author is not imparting medical advice. Before making major changes in your lifestyle or diet, please consult your medical care provider.

ENVIRONMENTAL IMPACT: Every book sold includes a donation to an environmental organization to help offset the carbon footprint of this book. This book's environmental impact estimates were made using the Environmental Paper Network Paper Calculator Version 4.0. For more information, visit www.papercalculator.org.

CONTENTS

PREFACE

This book challenges long-held beliefs and uncovers unpleasant truths, but it's ultimately a book of answers and hope; ideally, it's a constant reminder that a small yet crucial change can make a profound difference!

I recently heard someone say to "remember that out of the Dark Ages came the Renaissance"—a reminder to keep hope alive no matter the challenges, because good can come from all things. Those are the thoughts I hope you'll keep in mind as you read the following pages.

But for now, I'd like to take you on a short journey of the imagination. I would like for you to picture a dystopian world where one life form oppresses and murders countless others to satisfy their unquenchable tastes and pleasures. These self-serving activities, built on a foundation of suffering, are pursued vigorously and on such a massive scale that they threaten to disrupt the planet's ability to sustain life.

The oppressive business leaders and government officials from all walks of life are cultlike in their indoctrination of the masses into their ways. They preach the necessity of consuming their products, and now the populace is convinced they need them, which accelerates global demand with each passing year. As a result, government and corporate cult leaders churn out products with little regard for individual or planetary health and sustainability.

Operating in the shadowy recesses of governments and closed-door meetings, these government and corporate overlords profit while ignoring the harm and suffering they're creating. As a result, the truth is hidden from the masses, and the rich and powerful remain in control even though the systems they have built slowly begin to crumble, bringing the world down with them.

However, there's hope. A few rebels, having liberated themselves from the elite's grip, work to free all others. One by one, an army of resistance fighters begins to unite. Now the world is at a precipice, but the government and corporate leaders who brought it to the brink fight back relentlessly to discredit the rebels while ignoring the pleas of the oppressed. How will the story end?

What I've just described sounds like a work of fiction, and I wish it were. But unfortunately, most don't realize this dystopian world exists today; in fact, this book will prove we're living within it.

Animal rights activist and plant-based enthusiast James Aspey often cites during his presentations the following quote from Albert Einstein: "Those who have the privilege of knowing have the duty to act." I've seen firsthand the impact plant-based eating can have, but it wasn't until researching this book that I realized just how many more positive reasons there are for eating only plants. Too many scientific studies support plant-based eating to include them all, and the mountain of evidence is growing exponentially. Nevertheless, I've included what I consider the most significant ones while hopefully keeping this book a manageable and enjoyable read.

My research led me to the realization that we're at a critical time in our world. Alarms are going off in every direction—climate change, pandemics, the environment, food scarcity, health (including heart disease, diabetes, and Alzheimer's disease, among others), mass extinctions, wildfires, water shortages, and more. It feels like we're on a fast train to personal and planetary disaster, and scientists tell us time is running short if we're to turn it around.

But there's hope! Plant-based eating addresses many of the planet's most significant problems head-on, and I hope after reading this book, you'll agree so you, too, will escape the Meatrix, eat plants, feel great, and save the planet!

My spiritual teacher, Samuel, says that we change the world one by one by one. So if this book inspires one person to ditch animal products, my efforts will have been rewarded because I know there's nothing like the great feeling you get when you go plant-based. It's so much better for your health, the health of the planet, and the health of the animals.

My goal with this book is not to blame and shame people into becoming plant-based. It's to show why doing so is powerful and beneficial. I don't know a single person who has been plant-based their entire life, although I'm sure some do exist. For this reason, few of us can point a finger of blame at meat eaters and vegetarians for their lifestyles. We should not shame them. I'm privileged to have experienced firsthand the liberating feeling of living a life powered by plants and, therefore, feel compelled to act.

Another reason I'm against blaming and shaming others is that attacks are not an effective way to spread the plant-based message. Some describe vegans as self-righteous, extreme, and angry. I can understand why vegans might speak out in ways that could seem offensive. If they're like me, they've chosen to no longer participate in animal cruelty and are horrified by the needless suffering of animals. Most people have experienced the relief that comes when awakening from a bad dream. Plant-based eaters experience the exact opposite. They swallow the "red pill" and go from living a dream to waking in a nightmare of needless death and destruction. No wonder so many vegans feel betrayed and angry. The world we knew no longer exists, and it's a shock. Also, once you find something as liberating as plant-based eating, you want to share it with everyone. The problem is that most people don't want to hear it. I've found many people are uncomfortable discussing the food on their plates, and I don't think plant-based eaters will win many people over by making them feel defensive about their food choices. Based on the labels people use to describe vegans, it's no wonder so many people aren't interested in joining the plant-based revolution—who wants to join a group of self-righteous snobs?

Plant-based eaters need to do better at making veganism more attractive. Vegans are some of the most caring, loving people I know. It's too bad that these adjectives are not on most people's lips as they describe plant-based eaters. I look forward to a day when the world thinks of those who have escaped the Meatrix as intelligent, inspiring, and compassionate human beings whom others wish to emulate.

INTRODUCTION

As I write this, I am a fifty-nine-year-old man living in the United States. Sadly, I live in one of the country's poorer, least educated, unhealthy, and more addicted areas; however, I also live in one of the most beautiful places on earth filled with genuinely kind people—Lexington, Kentucky. Both sides of my family have been Kentuckians for generations, and no matter where I travel, Kentucky is always home to me.

Recent national data from the United Health Foundation shows that in 2021 Kentucky ranked forty-eighth for health behaviors and forty-seventh in health outcomes.[1] Only one state, West Virginia, had more residents with multiple chronic conditions.[2] In addition, as of 2020, the Centers for Disease Control and Prevention (CDC) ranked Kentucky's mortality rates in the top ten nationally in the following categories: cancer (one)[3], drug overdose (two)[4], septicemia (three)[5], chronic lower respiratory disease (five)[6], heart disease (eight)[7], kidney disease (ten).[8]

In fact, chronic conditions were so commonplace in Kentucky that I grew up never knowing options for avoiding or improving them existed. Instead, I thought these conditions were simply something everyone would eventually have to deal with and try to manage as best they could with pills, shots, or surgeries.

For example, when my mother's parents visited, my grand-father—who at the time was probably not much older than I am now and a stomach cancer survivor—traveled with multiple medications. Every day I watched him measure out the medicine for the nebulizer treatments he required three times a day to manage his asthma and emphysema. And over the years, I watched my grandmother's posture worsen as she became ridden with osteoporosis. Other relatives died of cancer due to chronic smoking, a common thing in a state known for its tobacco crops. And I watched members of my family yo-yo diet in an attempt to look better, feel better, and become healthier.

Finally (and thankfully), as a young adult, I decided to make some lifestyle changes that I hoped would offer me the best chance of avoiding some of those illnesses I saw in so many of those around me. For example, in 1985 (when I was twenty-three), I stopped eating meat and started exercising. Then in 2008, I took my health to a whole new level by going entirely plant-based, and I began running.

Some people will tell you that giving up animal products is foolish and will eventually wreck your health. But my life experience confirms what the science in this book will say: that the exact opposite is true.

Looking back, I think my decision to give up meat—and eventually all animal products—played a considerable role in my overall good health. But it's not just my opinion; my primary care physician also credits my excellent health to my lifestyle.

From the first time I stepped into his office at the age of fifty until today, my doctor and his residents have noted my excellent health. Practicing medicine in an area with such elevated rates of chronic disease, the doctors at the most highly rated healthcare system in the state, the University of Kentucky, were surprised when a fifty-year-old man walked into their office in vibrant health.

In the four routine physicals I've had in the last ten years, doctors have told me things such as, "We can't find any evidence you're actually in your fifties," and most recently (December 2021), "You're the healthiest person I've seen walk into this clinic." I don't say these things to brag—only to point out that I live in the same geographic area as many of my current doctor's other patients. Also, while I'm aware that grave inequities in access to healthcare do exist in the US, I use the same public healthcare system as many others in the bluegrass area of our state. So I'm more or less breathing the same air and drinking the same water, but the one major thing I'm not doing is eating the same food as most of my doctor's other patients. I think this is why he says I'm an outlier in his practice—not because I have lucky genes but because I take such good care of myself.

Speaking of genes, in case you're wondering, my gene pool isn't stellar. Only going back as far as my grandparents, my genetic inheritance is a who's who of medical conditions. Known family illnesses include asthma, autoimmune disease, breast cancer, cancer of unknown origin (which is a thing), emphysema, high cholesterol, hypertension, osteoporosis, stomach cancer, thrombosis, and weight issues (under and over). The great news is that according to Dr. Sharon Bergquist, the phytochemicals found in plants can change our DNA. In addition to their anti-inflammatory and antioxidant benefits, plant-based foods also play a role in gene expression—whether or not a "bad" gene turns on or off. Apparently, our genes do not need to be our destiny, which I believe is probably a relief to everyone. Lifestyle and the foods we eat play a more significant role than our DNA does in determining our health. Dr. Bergquist also states, "An 'epigenetic diet' is eating for the health of your DNA. Most healthful epigenetic foods discovered so far contain polyphenols, bioactive phytochemicals present in fruits, vegetables, seeds, and nuts."[9]

Most importantly, what this book discusses isn't rocket science. It's actually quite simple: eat plants and engage in moderate exercise. While these are two simple changes almost anyone can make, they have a profound impact not only on your health but the health of the animals and of the planet as well.

For instance, what surprised me most in researching this book, beyond the many health benefits of eating only plants, was the incredible impact of a plant-based lifestyle on the earth. Thankfully, more and more people are waking up to the knowledge that we *need* to do more to mitigate the worst-case scenarios of climate change. This book will hopefully prove to you that a plant-based lifestyle can not only help you avoid, slow down, or reverse most chronic illnesses plaguing Western societies, but it's also the most impactful thing we as individuals can do to prevent a global, catastrophic disaster.

How about you? Do you want more energy? Do you want to prevent, reverse, stabilize, or perhaps see a chronic health condition you're currently living with disappear? Are you concerned about climate change, and do you want to help the planet but don't know how? Maybe you're concerned with animal welfare, biodiversity loss, extinction rates, and emerging infectious diseases, and you want to find out more. If you're interested in any of the above, keep reading. The information on the following pages will empower you to make happier, healthier choices not only for yourself but for the entire planet. I encourage you to read on, take the red pill, and escape the Meatrix. If I can do it, I know you can do it too!

CHAPTER 1

What Is the Meatrix?

I've long thought that *The Matrix* movies were allegorical to our world and promoted a plant-based message. Whether the Wachowskis intended it as such, I don't know. Still, the idea that one life form—it's artificial intelligence in the movie—would use another life form for energy seems eerily familiar. We, humans, have used farm animals as our energy source. These animals, like the humans in *The Matrix*, have no control over their lives whatsoever. Every decision is made for them, including whether to reproduce.

Here's a quick synopsis for those who haven't seen the movies or don't remember the details.

The Matrix takes place in a dystopian future where humans are unwitting participants in a simulated reality created by artificial intelligence that has taken over the world and uses human bodies as an energy source. A computer programmer named Neo is offered a red pill to escape the Matrix and learn the truth about his world, or a blue pill to continue with his life as-is. Of course, Neo chooses the red pill.

The Meatrix differs from the Matrix in one critical way: instead of providing farmed animals an alternate—a virtual reality where they are oblivious of the actual conditions of their lives—farm animals are all too aware of their horrible existence. The Meatrix routinely separates farmed animals from their mothers, mutilates them without anesthesia, and raises them in cages or overcrowded conditions. Then they're exploited for their milk, eggs, and meat; injected with growth hormones and antibiotics; forcibly impregnated; and murdered at a very early age when they are no longer profitable to the Meatrix.

Some may argue that not all farm animals live bleak lives on factory farms. Still, data from the most recent USDA census of agriculture[10] (taken every five years) shows that factory farms account for 70.4 percent of cows, 98.3 percent of pigs, 99.8 percent of turkeys, and over 99.9 percent of chickens raised for meat, as well as 98.2 percent of chickens raised for eggs. Even if we lived in a world where all farmed animals were free to roam the countryside and graze in open pastures, at the end of the day, they would still be slaughtered for food—food that we don't need, would be better off not having, and that is killing our planet.

The Meatrix, like the Matrix, has created an artificial reality in which everything seems normal and acceptable. We go about our lives blind to the suffering of billions of feeling, intelligent animals and the damage eating them does to our health and the health of our planet. The Meatrix exploits innocent animals (instead of humans—which are also animals) as an energy source. In both the Matrix and the Meatrix, one life form (artificial intelligence and humans, respectively) imprisons the many. Both scenarios are insidious, but one is fiction, and the other is far too real.

Not only are the animals within the Meatrix powerless over the conditions of their lives, many employees within the Meatrix are similarly powerless over the conditions of their work environment. During the COVID-19 pandemic, many were unable to socially distance themselves and remain safe in the workplace. For example, an exposé from *The Guardian* details the callousness with which the Meatrix treats its employees.[11] Early in the pandemic, meat processing workers were forced to risk their lives to keep meat flowing to America's plates. In several newspaper advertisements, John Tyson, heir and chairman of Tyson Foods, described his business as being "as essential as healthcare." This book will clearly demonstrate that eating meat is not as essential as healthcare—or even healthy.

On May 12, 2022, Bloomberg News reported the findings of a House panel studying the nation's response to the pandemic. It revealed "a coordinated campaign by major meatpacking companies and their Washington lobbyists to enlist senior officials of then-President Donald Trump's administration in an effort to circumvent state and local health departments' attempts to control the spread of the virus in meatpacking facilities."[12] Obviously, their coordinated effort was successful and days after Tyson's newspaper ads, then-President Trump issued an executive order compelling meat-processing plants to reopen— an act applauded by the Meatrix.

In the following pages, I reveal some of the many ways the Meatrix exploits us with its products. But *The Guardian's* exposé reveals just how thoroughly the Meatrix exploits even its own workers. And it's not only Tyson Foods. Kenneth Sullivan, CEO of Smithfield Foods, an international pork producer overseeing more than 2,500 farms with over fifty thousand employees and a reported annual revenue of $14 billion, labeled the workers' deaths during the first two months of the pandemic as a "culture issue." According to documents released by *ProPublica*, Mr. Sullivan wrote, "Social distancing is a nicety that makes sense only for people with laptops."[13]

As a result of being compelled to work in conditions even the business owners said were unsafe, forty-nine meat-processing workers died of COVID-19 within the first few months of the pandemic. Sadly, outbreaks and deaths weren't confined to the factory. Workers took the virus out of the crowded meatpacking facilities and into their communities. "Researchers found that by July 2020, areas nearby meat-packing plants had far more COVID cases and deaths than expected: about 5,000 additional COVID deaths and about a quarter-million additional cases." In all, 6 to 8 percent of all early cases of COVID-19 and 3 to 4 percent of all early COVID-19 deaths in the US were connected to meat-packing plants.[14]

"This needs to go down in the history books as one of the biggest failures to the working man or woman that this country's ever seen," said Kim Cordova, president of the United Food and Commercial Works Local 7, the union representing meatpacking workers at the Greely, Colorado, factory owned by meat conglomerate JBS.[15]

Clearly, defining meat plant employees as "essential workers" came at a considerable cost of lives lost. While this book proves the work meatpacking employees do is *not* essential to our health but actually detrimental to it, the workers themselves, of course, *are* essential people and should not be so callously treated as expendable.

Like in *The Matrix*, there are rebels living outside the Meatrix who fight to free others so they can live happier, longer, more productive lives. I'm one of those rebels, and this book is your red pill that will awaken you to the truth about the Meatrix. Of course, if you prefer the blue pill and want to continue with life as it is, then you may as well stop reading now. But if you make that choice, you'll be robbing yourself of the great gift of a healthier, more compassionate life and continuing on a path that spells disaster for your health and the planet.

Simply put, the Meatrix is the world into which we were all born. It's like a computer's operating system, constantly running in the background of our devices and therefore nearly invisible and rarely questioned. None of us woke up one morning saying, "I think I'll try eating an animal today," or "You know, I feel like imprisoning mother cows and drinking their babies' milk!" No. Few if any of us consciously chose to eat animal products. And because meat-eating is all we've known since an early age, most of us never question it. Most people don't even realize the choice to consume animal products was made for us by our parents and their parents before them, and the fact that it's always been that way in our families or our lives doesn't mean it's good or can't be changed.

The tentacles of the Meatrix are many and far-reaching. I remember seeing posters in my elementary school's cafeteria promoting the power of milk and the importance of meat. What other industry has access to advertising to young children in schools? We shouldn't underestimate the power of the meat and dairy industries.

We just learned how meatpacking employees were forced to risk their lives to put meat on our plates early in the pandemic. It might sound cynical, but I believe meat conglomerates like Tyson Foods, Smithfield Foods, and JBS made strategic moves so they could wield such power over their workers' lives. Years ago, meatpacking workers were mostly American-born and earned the equivalent of twenty-five dollars per hour in today's economy. Beginning in the 1970s, meat conglomerates intentionally moved their operations out of urban areas and into rural America, where jobs were scarce. By migrating their operations from cities to the countryside, meat companies became the single largest employer in these communities. Also, as a result of this move, the demographics of meatpacking plants saw a dramatic shift from American-born to immigrant employees—a strategic move allowing meat companies to exploit a demographic with little collective bargaining power.[16]

When I was researching this book, it became clear just how

powerful the Meatrix is. Seeing how a meat company's ads in newspapers could influence a president to compel employees back to work illustrates the cozy relationship the Meatrix has with our government. I was reminded of Dwight D. Eisenhower's farewell speech to the nation on January 17, 1961, in which he warned the American public of the growing influence and grave implications of a permanent armament industry known as MIC—the military–industrial complex. Unfortunately, today I believe we're living within another just as powerful MIC conglomerate: the meat–industrial complex. The meat–industrial complex is a vast network of governmental institutions, influential councils, pseudo-government boards, powerful corporations, industry insiders, and well-funded corporate lobbyists. These entities conspire to squelch and, if that's not possible, discredit all critics. They work to make the Meatrix as profitable as possible, while people's health and the health of the planet go down the toilet.

Yet another example of the meat–industrial complex's power is shared in Marta Zaraska's book *Meathooked*:

> On June 3, 2013, a seemingly trivial sign appeared at one of the food stations in the white expanse of the Longworth cafeteria—the bright, open space located in the same building as the offices of the House Agriculture Committee in Washington, D.C. It's here on workdays around noon that a long line of staffers forms: members of Congress and lobbyists await their turn to grab lunch. On that particular Monday, many of them spotted a sign advertising one of the food options. The sign, supposedly placed by one of the cafeteria's employees, simply said: "Meatless Monday."[17]

This sign, promoting just one out of many options offered by the cafeteria, created quite a stir within the Meatrix—so much so that on June 7, the Farm Animal Coalition (an organization that includes some of the nation's most prominent ranchers and farmers and is part of the meat–industrial complex) issued a statement to the House Administration Committee that included the following: "'Meatless Mondays' is an acknowledged tool of animal rights and environmental organizations who seek to publicly denigrate US livestock and poultry production." As a result, on June 10, the following Monday, the "Meatless Monday" sign at the Longworth Café was no longer displayed and has not reappeared.[18]

The Longworth Café incident is just one example of the power of the Meatrix. All you have to do is follow the money to understand how influential the Meatrix is in our society. According to the North American Meat Institute, the US meat and poultry industry generated an estimated $1.2 trillion of economic activity in 2019.[19] By comparison, in 2020, the total US sales of plant-based meat and dairy alternatives, totaling $7 billion, was 120 times less than meat and poultry.[20]

As part of the meat–industrial complex, the United States Department of Agriculture (USDA) determines our country's nutritional guidelines. But according to the USDA Strategic Goals 2018–2022, its number one goal is to "ensure USDA programs are delivered efficiently, effectively, with integrity and a focus on customer service." Second on their list is to "maximize the ability of American agricultural producers to prosper by feeding and clothing the world." The last item on their list of strategic goals is to "provide all Americans access to a safe, nutritious, and secure food supply." Clearly, the USDA's primary concern is *not* nutrition and our health.[21]

The fact that the same agency that represents cattlemen and factory farmers creates our country's nutritional guidelines is a conflict of interest. Wouldn't it be better if an independent group of nutritional scientists made the recommendations for our dietary guidelines instead of one with economic ties to the Meatrix? Because of this conflict, the US nutritional guidelines contain animal products that science has shown to be harmful to our health. If you want a sneak peek, Chapter 11 discusses the USDA dietary guidelines in greater detail.

Here's another example of how prevalent the Meatrix is in our society. I recently accompanied my partner, Lea, to her doctor's office for a bone density test. They asked her to complete a questionnaire ahead of time. One question asked how many times a week she consumed cheese, yogurt, and milk. If you ask just about anyone on the street why we are encouraged to drink milk, they will almost always say, "For strong bones," or "To get my calcium." The Meatrix has falsely convinced us to think that consuming dairy products leads to stronger bones. Whether or not this doctor's office is aware of it, their questionnaire is helping promote the false narrative that you need to consume dairy products to have strong and healthy bones.

That such a questionnaire exists today, despite mountains of evidence that contradicts dairy is good for strong bones, speaks to the power of the Meatrix.

In the following chapters, I share some of the overwhelming scientific evidence that the Meatrix contributes considerably to chronic diseases, reduced life spans, and climate change. Unfortunately, the Meatrix remains so powerful in part because it began indoctrinating us at a very early age, and now we find it hard to change our thoughts about it. But changing our beliefs *is* possible and necessary if we want to escape the Meatrix.

For example, as mentioned above, many people believe that consuming dairy products leads to stronger bones. Unfortunately, however, countries with the highest dairy intake also have the highest rates of osteoporosis, which is a significant problem in the US and other parts of the world.[22] But this fact seems counterintuitive simply because it contradicts the benefits the dairy industry claims in its ads dating all the way back to the 1980s' "Milk. It does a body good." campaign.

In fact, in 2014, *The BMJ* published a report based on two Swedish studies of more than one hundred thousand people with a follow-up of twenty-plus years. These studies concluded that milk doesn't do a body good: "High milk intake was associated with higher mortality in one cohort of women and in another cohort of men, and with higher fracture incidence in women."[23]

Additionally, *The Journal of the American Medical Association* reported the findings of the Nurses' Health Study covering ninety-six thousand white men and postmenopausal women aged fifty years and older from the Health Professionals Follow-Up Study in the United States. The study concluded, "Greater milk consumption during teenage years was not associated with a lower risk of hip fracture in older adults."[24]

So getting back to the doctor's questionnaire, if they wished to determine what Lea might be doing to have stronger bones, they could have asked how often she does weight-bearing exercise[25] or how many servings of dark, leafy greens she eats each week.[26]

As a person who has escaped the Meatrix, watching the plant-based movement grow heartens me. A 2020 Ipsos Retail Performance study revealed the number of people identifying as plant-based in the US was 9.7 million—an increase of over 300 percent in the last fifteen years.[27] It's not only a US phenomenon; the number of vegans in the UK quadrupled between 2014 and 2019.[28] As more plant-based foods are coming to market, more and more people are ditching the meat and opting for plants.

Kelly Fairchild, global business development manager at Ipsos Retail Performance, said:

Plant-based diets are fast becoming mainstream, but the change hasn't been a steady one. Recent years have seen rapid adoption of vegan diets and more meat-free products making their way onto shelves. As the dialog around veganism shifts from one of animal welfare, to wider concerns around climate change and personal health, we are seeing more and more people adopt this once minority dietary preference.[29]

The recent, exponential growth of plant-based eating gives those who have taken the red pill and escaped the Meatrix much to celebrate. For instance, stock in the plant-based meat company Beyond Meat soared 163 percent over its opening price on its first day of trading.[30] Since then, its products are gaining even more popularity and becoming less expensive.

According to Euromonitor, the global market for plant-based meat alternatives is expected to grow to $23.4 billion by 2024. In the last few years, the price of Impossible Foods plant-based meats has decreased 15–20 percent. Impossible Foods president Dennis Woodside said, "Currently, our product on shelf is priced a little higher than organic grass-fed beef, so that's still a premium price, and we know we need to get that down over time." While the Meatrix has fine-tuned its operations for decades, squeezing as much efficiency as possible out of its operations and enjoys enormous subsidies from the government, plant-based meat companies are much younger by comparison. It will likely take start-ups like Beyond Meat and Impossible Foods fifteen to twenty years to achieve parity with the Meatrix's prices.[31]

In 2019, in taste tests across the country, Burger King surprised many meat eaters who learned the Whopper they were enjoying was plant-based. Afterward, Burger King launched an Impossible Whopper (made with Impossible Foods' plant-based meat). As a result, the fast-food giant experienced a whopping 6 percent same-store sales growth rate (a metric used for financial analysis of a company's growth).[32]

In 2019, when testing plant-based boneless chicken wings and nuggets, Kentucky Fried Chicken (KFC) partnered with Beyond Meat and became the first national chicken chain to offer plant-based meat. The plant-based chicken wings were so popular they sold out at KFC's single test location in Atlanta in only five hours! Plus, sales of Beyond Chicken exceeded an entire week of KFC's best-selling popcorn chicken.

In addition to Burger King and KFC, Subway launched a Beyond Meatball sandwich. This sandwich proved so successful it became a permanent menu item. These companies' successes led other companies, such as McDonald's, to test plant-based meat in some of their markets. So now plant-based food is quickly becoming fast food!

In the summer of 2019, while I was in Atlanta on business, I heard about a new plant-based restaurant that had recently opened, the Slutty Vegan. My Atlanta friends told me lines formed outside this eatery regularly. So, eager to see it myself, I went to check it out that Saturday night. I arrived at 7:30 p.m. and saw a line of people extending down the entire block and around the corner—too long for me. But I chatted with a mother and her son whose position in line had them close to the restaurant's entrance. In speaking with them, I discovered they were not strict vegans, but they were vegetarians. I wondered how long these non-vegans had been waiting in line for a plant-based meal and asked. Their reply astonished me. They had been waiting in line for two hours!

Then in 2021 Beyond Meat partnered with the fast-food restaurant chain Panda Express to develop a plant-based twist to PE's most iconic dish, orange chicken. After unprecedented sales in New York and Los Angeles stores, it was announced in late October 2021 that the dish would be launched in seventy new US locations for a limited time.

Finally, in March 2022, something considered unimaginable only a few years ago happened when Burger King surprised many Londoners by announcing its marquee restaurant in Leicester Square was going 100 percent plant-based for one month. Critics complained that forcing everyone to eat plant-based foods was manipulative, but experts say fast food's reliance on processed meats, condiments, and cheese, makes it a perfect gateway plant-based food. Burger King spent years developing the menu which included twenty-five options. Katie Evans of Burger King UK commented that the menu "goes hand in hand with our target of a 50 percent meat-free menu by 2030, as well as our commitment to sustainability and responsible business. We can't think of a more fitting way to re-launch our new-look flagship in Leicester Square."[33] James Lewis, who is a marketer and product developer for the plant-based restaurant 123V on Bond Street, London, said, "Not too long from now, people will be getting their burger and it'll be a vegan one, and that'll be the norm and they won't think any different.... There's no point starting a vegan chain because once McDonald's figures out how to make a good vegan burger, they will think: 'What's the point in the cost of keeping all these animals when we can make it just as good and grow it in the ground?'"[34]

In addition to the recent changes seen in the fast-food industry, 2021 saw a growing number of plant-based IPOs such as Meatech, Oatly, Odd Burger Corporation, Yumy Bear, and Eat Well Group (formerly Eat Well Investment Group). Renub Research projected the US plant-based food industry alone will nearly double, growing from $5.6 billion in 2020 to $10.7 billion in 2027.[35]

For everyone who has escaped the Meatrix, these are all encouraging signs of a world that's changing. But it's important to point out that the way these products are served might not be entirely plant-based. For instance, Subway's Beyond Meatball sub isn't fully plant-based unless you omit the cheese and choose one of their vegan breads. And while these plant-based options are indications that companies are willing to test the waters with meat alternatives, many are still in the business of selling animal products, and that's not going to change until more people demand it.

How do I know fast food restaurants will become more plant-based if people demand it? I experienced such accommodations to dietary preferences firsthand while traveling in India in 2011. I was in Kolkata, and my guide took me to a newly opened McDonald's restaurant. Indian culture venerates cows, and therefore, many consider eating beef taboo. You may ask yourself, then, how a McDonald's restaurant could exist within such a culture. My guide was eager to show me. The McDonald's in Kolkata (and presumably elsewhere in India) is really two restaurants under one roof. There are separate lanes for ordering, and there are even separate kitchens—one that prepares meat and another that doesn't. It was fascinating to see how far McDonald's is willing to go to attract customers.

Sadly, despite the explosion of plant-based eating in the US and elsewhere, the Meatrix is growing at an astonishing rate around the globe—a fact that should be very alarming to every medical professional, climate change scientist, environmentalist, conservationist, and animal rights activist on the planet. For example, according to Our World in Data, global meat production quadrupled from 1961 to 2018, while the global population only increased roughly two and a half times. During this time span, Europe saw a twofold increase in meat production, while the US output increased 2.5-fold.[36]

However, these numbers pale in comparison to other parts of the world. In cultures where diets have been traditionally

MEAT PRODUCTION,1961–2018 (MILLIONS OF TONS/YEAR)

Meat includes cattle, poultry, sheep/mutton, goat, pigmeat, and wild game.

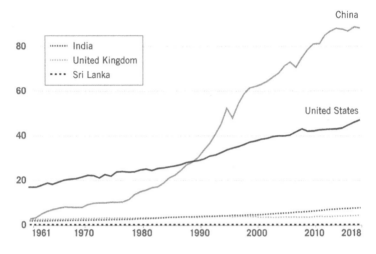

Source: Hannah Ritchie and Max Roser, "Meat and Dairy Production," *Our World in Data* (2017), ourworldindata.org/meat-production

plant-based, people are turning to meat at an alarming rate. For instance, according to a 2011 report, "Livestock is one of the fastest-growing sectors in agriculture, potentially presenting opportunities for economic growth and poverty reduction in rural areas, though unless carefully managed the main social effects may be negative." The report continues saying that in just thirty years, meat consumption in China rose from less than nineteen pounds per person per year to more than 110 pounds.[37]

Likewise, the report projects that by 2030, poultry consumption in India will increase by 1,277 percent from year 2000 levels. The same is happening all over Asia: by 2030, meat consumption will increase by 159 percent in Malaysia (beef), 146 percent in Cambodia (beef), and 1,049 percent in Laos (poultry). Meanwhile, China will consume over twenty-two million tons more pork! With a combined 2.7 billion people in India and China, this increased appetite and demand for meat spells disaster for our planet and the health of its people.

But what's looming beyond those alarming 2030 projections? Sadly, it only gets worse. The global population is projected to grow from 7.6 billion people to 9.8 billion people by 2050. These 2.3 billion additional people will demand more meat and animal-based foods, fueling declining health and even greater environmental damage to the earth. The demand for meat and dairy products is expected to increase more than 60 percent by 2050![38]

FASCINATING FACTS

- The US meat and poultry industry generated 1.2 trillion dollars in 2019.
- The poultry consumption in India is projected to increase by 1,277 percent from year 2000 levels by 2030.
- By 2030, China is projected to consume an additional 22,050,000 metric tons more pork per year than in 2000.
- The estimated value of all plant-based protein is projected to be 23.4 billion dollars in 2024.
- Global demand for meat and dairy will increase by 60 percent by 2050.

CHAPTER 2

Out of Sight, Out of Mind

The Meatrix is masterful at portraying a pastoral version of modern farm life while hiding the horrific reality of factories and smaller farms. Few people ever stop to consider the origins of their food. Most purchase the neatly packaged body parts of slaughtered animals and never stop to think about the animal whose body they're buying. And few seem concerned about the animal's quality of life before slaughter or the carbon footprint of raising and delivering that body part to them. Because we're so removed from the life cycle of our food, most of us blindly eat the same things our parents, school, or society put on our plates as children.

There's a story within cooking circles of a young woman who would remove the butt end of a roast before placing it in her pan with vegetables. When asked why she did that, she replied she had learned the technique from her mother. When her mother was asked, she also replied this was how her mother had taught her. When finally the grandmother was asked, she said she never had a pan large enough to accommodate a whole roast, so she just cut the end off.

The young woman was following what she thought was a rule of cooking, so she never asked questions and therefore never knew that the only reason for trimming the roast was because her grandmother's pans were too small. This story is a powerful reminder of how we blindly follow and repeat traditions passed down from our parents without thought. Unfortunately, it often takes something or someone to shake things up a bit to help us see the world from another perspective.

In the United States, pharmaceutical companies can advertise on television but must mention a drug's known side effects. So it's not surprising that these pharmaceutical ads show images of happy people living large with friends and family, while a narrator quickly reads through all the drug's known side effects. I've often wondered how impactful these ads would be if they instead presented the viewer with images of people experiencing the mentioned side effects. For example, if uncontrollable diarrhea is a known side effect, instead of watching a grandpa lifting his grandchild, we see him running to the nearest bathroom. Horrible marketing but honesty in advertising!

Pharmaceutical advertisements on American television are highly effective. As a result, patients today often ask doctors for a specific brand-name prescription. But it's not just direct to consumer (DTC) advertising. One reason pharmaceutical advertisements are so effective is drug companies target medical health professionals, too. In fact, of the $30 billion spent each year on pharmaceutical ads in the US, $20 billion is spent advertising to doctors and other medical professionals. It's a one-two advertising punch that leads to more prescriptions and higher medical costs.[39] The fact is, marketing works, period. Similarly, the Meatrix's relentless marketing of animal products highly influences the average consumer.

Like big pharma, the Meatrix doesn't want us thinking about the horrible "side effects" of eating its products, so it frequently avoids any imagery of farm animals, the environmental destruction it creates, or people who become ill from eating its products. When the Meatrix does use images of animals to promote products, it shows consumers happy cows on cartons of milk (like Elsie of Borden Dairy) and smiling pigs sporting sunglasses and leather jackets on packages of bacon (like Boss Hog). At our local Trader Joe's, there's a large mural depicting cows and chickens strutting on a racetrack for thoroughbred horses (this is Kentucky, after all), another industry built on animal abuse and exploitation. These cartoonish images perpetuate a total disconnect from the harsh reality that these animals face. The Meatrix knows that if people considered what occurred on most farms and in slaughterhouses, they would demand change.

When asked, almost every American says they're concerned about animal welfare, so the Meatrix tries to hide from the public what it's doing to animals. In fact, a 2012 poll conducted by Lake Research Partners revealed that 94 percent of the public agrees that "from every step of their lives on a farm—from birth to slaughter—farm animals should be treated in a way that inflicts the least amount of pain and suffering possible." The study also concluded that 71 percent support undercover investigative efforts to expose animal abuse on industrial farms.[40] Likewise, in a 2018 survey conducted by the American Society for the Prevention of Cruelty to Animals (ASPCA), 76 percent of consumers across every demographic expressed concern toward animals raised for food.[41] The Meatrix knows that an overwhelming number of people are against many of the practices required when treating farm animals as food commodities, so it works hard to keep the public from seeing the dark underside of the industry.

In fact, as of June 2021, ASPCA identified six states (Alabama, Arkansas, Iowa, Missouri, Montana, and North Dakota) with "ag-gag" laws that criminalize attempts by whistleblowers to document the conditions of factory farms. Fortunately, despite the Meatrix's attempts, ag-gag legislation has been defeated in nineteen states and found unconstitutional in five.[42]

Kentucky tried to pass its own "ag-gag" law in 2014 by attaching it to a pro-animal rights bill.[43] Thankfully, the bill died, but I'm sure the Meatrix will continue to try to get its way in Kentucky and elsewhere.

In 2019, I had the fortune of attending a talk at Morehead State University in Eastern Kentucky by animal rights activist Jenny Brown, co-founder of the Woodstock Animal Sanctuary in New York. I have vivid memories of the evening. When I walked into the theater, the talk was already underway. I was surprised and delighted by the many people in the audience. I know that veganism is growing in popularity among young people and on college campuses, so I assumed that accounted for the robust turnout.

Jenny Brown no longer works at the Woodstock Animal Sanctuary but gives animal rights presentations to philosophy students on college and university campuses across the US. When she transitioned into the question-and-answer portion of the evening, it became clear that not everyone in attendance was studying philosophy. It turned out students and faculty in the university's agriculture department had caught wind of the presentation and attended in numbers seemingly greater than those of the philosophy students.

During the contentious Q&A session, an agriculture student's response to the suffering farmed animals face was that all industries have their problems. Of course, all industries do have problems, but I can think of no other that exploits and kills so many unwitting participants. Undoubtedly, the Meatrix has more than its share of problems.

A different ag student mentioned insulin and estrogen, which she said were significant Meatrix contributions to society. These connections were unclear to me, so I researched to learn more the next day. I discovered that in the 1980s, recombinant human insulin, synthesized by inserting human insulin into *E. coli* or yeast, replaced insulin derived from animals (pigs and cattle). Eli Lilly's "Humulin" went on sale in 1982, essentially putting an end to insulin derived from nonhuman animals.[44]

The second item, estrogen, found in the hormone-replacement drug Premarin, is still derived from animals. Mares are forcibly impregnated and confined in cramped stalls so big pharma can collect their estrogen-rich urine.[45] However, just like with insulin, vitamin B12, omega-3, and vitamin D supplements, plant-based, cruelty-free estrogen options are available. So this student's information was outdated.

Researching this book has brought me a deeper understanding of the harm the Meatrix does to the planet, animals, and those who consume its products. But that night in Morehead, Kentucky, I came to a deeper, more nuanced understanding of the plight of today's farmers. I touch on this a bit in Chapter 10 on cognitive dissonance, but I now see that farming isn't so much a job as a vocation. It's human nature to defend one's actions, even when we're confronted in a loving and compassionate way, as was the case during Ms. Brown's presentation that night. If I discovered generations of my family had harmed animals and helped destroy our planet, instinctually my reaction would likely be to become defensive and disagree.

I later wrote to Ms. Brown and asked her thoughts on the evening, and she replied that the meeting "was an example of the defensiveness that is often seen when deeply held cultural beliefs are challenged. People want to remain comfortably numb and unaware of what they're paying for so they don't have to change or feel bad about their eating habits."

Through its marketing slogans, the Meatrix tries to hide the reality of what people are paying for so they'll continue to "live comfortably numb and unaware." "Free-range," "cage-free," "cruelty-free," "grass-fed," "welfare meat," and "humane slaughter" are marketing soundbites meant to pacify those within the Meatrix. However, research and investigations have shown that these labels reflect little about the experience of the animals.

For example, the "free-range" label invokes images of animals roaming free on green pastures with plenty of sunshine and fresh air. However, a 2014 article by Certified Humane reports the "USDA considers five minutes of open-air access each day to be adequate for it to approve use of the 'Free Range' claim on a poultry product." Additionally, while the USDA's definition for "free range" indicates birds must have either "outdoor access" or "access to the outdoors," access doesn't need to be full-body. For instance, the USDA defines birds with a "pop hole" to the outdoors as "free range."[46]

Sadly, in the case of chickens bred for meat (instead of eggs), free-range means virtually nothing, as these chickens are drugged and bred to grow so obese they can hardly move, let alone go for an outdoor stroll. Richard Lobb, the spokesperson for the National Chicken Council, said, "If you go to a free-range farm and expect to see a bunch of chickens galloping around in pastures, you're kidding yourself."[47] Yet this is precisely the image I believe the Meatrix intends to create in the minds of consumers.

Let's look at another familiar Meatrix phrase, "humane slaughter." Humane means having or showing compassion. I would say that humane slaughter is an oxymoron because there is no way to compassionately take the life of a being that doesn't want to die. It's simply impossible to rob someone (and an animal is a someone, not a thing—see Chapter 4) of its most precious gift—life—and call it humane. This illustration is extreme, but let's say I was a great husband to my wife and treated her well. We went on vacations together, celebrated our anniversaries, and had a great life together. What if one night I killed her quickly while she lay sleeping? Could I go to the judge and say, "It's okay, Your Honor. It was a humane slaughter"? No way. The judge would say it didn't matter if she suffered or not; it was not humane because I ended her life—a life she wished to continue living. So how can we say killing anyone, including farmed animals, is humane if they don't want to die?

As mentioned earlier, polls show that most people believe that we should do no unnecessary harm to animals. If you agree, you are already plant-based by your values because breeding, raising, and killing animals does immeasurable harm and is not necessary in today's world. The millions of plant-based people worldwide are proof of that. The only thing required for you to escape the Meatrix is the most vital step of aligning your actions with your values. If you believe it's wrong to do unnecessary harm to animals, then you owe it to yourself, the planet, and the animals to act in a way consistent with your beliefs by becoming plant-based. The great news is it's never been easier.

FASCINATING FACTS

- 94 percent of Americans are concerned about animal welfare.
- 76 percent of consumers across every demographic expressed concern toward animals raised for food.
- Only 5 minutes a day with access to the outdoors qualifies for "free range."
- Six states have "ag-gag" laws that criminalize attempts by whistleblowers to document the conditions of factory farms.

CHAPTER 3

Escape the Meatrix for Your Health

The data in this book is something the Meatrix doesn't want you to know, and it works to keep this information hidden from you so you will continue to purchase their products. As you'll see, the evidence presented in the following pages affects not only your health but the health of our planet because it's becoming alarmingly clear that is what is at stake.

As I mentioned in Chapter 1, most people on the planet spend their entire lives living inside the Meatrix without ever consciously choosing to do so. My family introduced me to the Meatrix at a very early age. As a result, eating meat was a mindless act, similar to the young woman who cut off the end of her roast without understanding why. We repeat habits and ways of eating learned in childhood without considering all the costs involved.

However, there are many reasons people like me choose to escape the Meatrix. For some, it's about the animals; for others, they are more concerned with climate change and the environmental impact. Still others choose to live outside the Meatrix to improve their health.

Generally speaking, there are seven reasons for choosing to live outside the Meatrix. I will discuss each of these in the following three chapters. They are: health, pandemics, and emerging infectious diseases; climate change and the environment; and treatment of animals, humans, and spirituality.

HEALTH

The Meatrix is big pharma's best friend because eating within the Meatrix makes us sick. Yes, that is a bold statement that this section proves to be true. The Meatrix fuels many chronic diseases that plague countless Americans, and big pharma cashes in by advertising drugs to address these health conditions. This may sound cynical, but it's a vicious cycle, and it's why I say we don't have a healthcare system in the US, but a sick-care system.

In his bestselling book *Fiber Fueled*, gastroenterologist Dr. Will Bulsiewicz explains the powerful impact food has on our health in the following way: the average American eats three pounds of food a day, or roughly one thousand pounds each year. Therefore, throughout a lifetime, the average American eats approximately eighty thousand pounds of food. Unfortunately, for most people, these eighty thousand pounds are composed of the standard American diet, full of animal-based foods proven to be detrimental to our health. Dr. Bulsiewicz says this unhealthy diet often leads to a common symptomology of tiredness, body aches, poor digestion, depression, and a litany of chronic diseases. Unfortunately, Western medicine attempts to counteract eighty thousand pounds of an unhealthy diet with isolated compounds in pharmaceuticals and supplements rather than addressing the root cause: the diet itself.[48]

But wouldn't it be better if we never contracted these chronic conditions in the first place? As this section will prove, plant-based lifestyles offer protection from, slow down, and even reverse many of the chronic conditions our sick-care system is frantically trying to address.

The following graph from the Peterson-KFF Health System Tracker lists 2021's top eleven causes of death daily in the US. In this chapter, I'm going to present scientific evidence that a plant-based lifestyle can protect you from nine of the top ten leading causes of US deaths. These are not one-off studies, but part of a growing mountain of evidence—all pointing to a powerful and singular choice: a plant-based lifestyle as a potent antidote to the dangerous side effects of our typical Western dietary pattern.

Since joining the plant-based movement in 2008, and based on my research and experiences, there's no doubt in my mind that a well-balanced, whole-foods, plant-based lifestyle improves one's health significantly. We've all seen the recommendations to eat more fruits and vegetables and fewer animal products, and it makes sense. Fruits and vegetables have many benefits, such as the vitamins and minerals needed for a healthy body.

In fact, LiveScience reports that the United States Department of Agriculture "notes that consuming a phytonutrient-rich diet seems to be an 'effective strategy' for reducing cancer and heart disease risks." Phytonutrients are chemical compounds produced by plants that have antioxidants for combating free radicals and anti-inflammatory benefits. The article also reports that phytonutrients may enhance immunity and improve intercellular communication.[49]

Plant-based foods also provide dietary fiber, which offers benefits "such as helping to maintain a healthy weight and lowering your risk of diabetes, heart disease, and some types of cancer," according to the Mayo Clinic.[50]

I'll now go through the list of the top eleven daily causes of death in 2021 and share research on how a plant-based lifestyle can help prevent suffering from these chronic conditions.

AVERAGE DAILY DEATHS IN THE U.S. FROM COVID-19 (AUGUST 2021) AND OTHER LEADING CAUSES (2021)

1. Heart disease
2,050

2. Cancer
1,621

3. COVID-19
727

4. Accidents (unintentional injuries)
553

5. Stroke
440

6. Chronic lower respiratory disease
374

7. Alzheimer's disease
332

8. Diabetes
275

9. Other diseases of the respiratory system
180

10. Renal failure
144

11. Suicide
126

Notes: The COVID-19 daily death average is for August 1 to August 24, 2021, and is based on KFF COVID-19 tracker data. Accidents and suicide daily death averages are for 2020. The Alzheimer's disease death average is calculated from the first day of January 2021 to the last day of June 2021. Average daily deaths for all other leading causes are from the CDC and are from the beginning of 2021 to the last MMWR week of June 2021.

Source: KFF analysis of CDC mortality and KFF COVID-19 tracker data.

1. Heart Disease

In addition to suggesting we eat more fruits and vegetables, nutritionists often recommend reducing our consumption of saturated fats found in butter, cheese, meat, and other animal-based foods. For instance, research presented by the American Heart Association found that plants with monounsaturated fats—avocados, nuts, seeds, and vegetable oils—are associated with a lower risk of dying from heart disease and other causes, while saturated fats found in animal products increased that risk.[51]

The research followed over ninety-three thousand people for twenty-two years and looked for patterns (not causes and effects). Marta Guasch-Ferre, one of the lead researchers in this project, said, "We have observed a beneficial role of monounsaturated fats for the prevention of cardiovascular and total mortality when plant-based foods are the primary sources."[52]

Thinking of the decreased mortality rate of those eating primarily plant-based monounsaturated fats reminds me of a conversation I had at my gym with Ted, a lifelong runner. He had recently had a heart attack and received treatment typical in the US: a stent and four medications. Curious, I asked if his doctor suggested any dietary changes, and he said no. I about fell over. I thought the doctor would have at least said something about the importance of limiting saturated fats. In today's world, with all the information now available linking the consumption of animal products to heart disease, I would call that malpractice.

Heart disease and cancer are the number one and two causes of death in the US, and together they account for 46 percent of all deaths.[53] A quote attributed to Hippocrates, the father of modern medicine, is, "Let food be thy medicine and medicine thy food." Generally speaking, people seem to fall into the "food as thy medicine" camp or the "medicine as thy medicine" camp. Ted's doctor was obviously in the latter.

Unfortunately, Ted's doctor might not be an outlier. Despite overwhelming evidence that consuming animal products can be harmful to one's health, few doctors ask their patients about what they eat. Researchers in the late 1990s determined that doctors failed to discuss diet and nutrition with their patients in a "comprehensive and effective manner"; unfortunately, little has changed in the past twenty years, as doctors continue to report that they rarely have conversations about diet and nutrition with their patients.[54]

One reason physicians rarely broach nutrition with their patients may be due to medical students receiving very little training in diet and nutrition. According to an article published by *US News and World Report*, medical students receive an average of only 19.6 hours of dietary education spread across four years of training.[55] However, "nutrition is a cornerstone of cardiovascular health," according to a 2017 study published in the journal *Current Cardiology Reports*, "yet the training of cardiovascular specialists in nutrition has been called into question."[56]

In *Fiber Fueled*, Dr. Will Bulsiewicz writes that most medical students spend months learning "the nuances of drug pharmacology, but formal nutrition training can be just two weeks or less." For Dr. Bulsiewicz, his nutritional training occurred in his second year of medical school. He wouldn't complete his training to be a licensed gastroenterologist for another ten years. And, during those ten years, his training never included nutrition again. He further explains that even if he provides nutritional counseling to his patients, as a gastroenterologist, he's unable to bill for the time he spends doing so. As a result, he says, "Our system penalizes doctors for taking the time to discuss nutrition."[57]

The critical role nutrition plays in health is further confounded when specific medical journals offer advice contrary to the evidence. A blatant example of this occurred in the *Annals of Internal Medicine* in 2019, when researchers recommended that consumers continue to eat red and processed meat at their current levels. The authors' recommendation was not based on the peer-reviewed science cited in their study, but rather on human nature, saying that people derive pleasure from eating meat and are unlikely to change their behavior. It's worth noting the study referenced many articles with research funded by the beef and pork industries.[58]

The Physicians Committee for Responsible Medicine questioned whether the "*Annals* articles are mere clickbait, published in anticipation of a media frenzy,"[59] and Walter C. Willett, professor of epidemiology and nutrition at Harvard's T. H. Chan School of Public Health and professor of medicine at Harvard Medical School, said of the *Annals* article, "It's the most egregious abuse of data I've ever seen."[60]

I believe what we eat has significant impacts on our overall well-being, and whole-foods, plant-based eating is a powerful medicine for the body and soul. But you don't need to take my word for that. The *European Journal of Pharmacology* published an article that concluded that "increasing evidence shows that, particularly during the early stages of disease, sustained lifestyle changes are by no means inferior to drug treatment, and often even more efficacious in stabilizing or even reversing the disorder."[61]

Thinking of "food as thy medicine" reminds me of something that recently happened in my partner's life. She has lupus, and her doctors monitor the progression of the disease through bloodwork. Recent labs indicated her potassium levels were low. After saying she didn't want to go on one more medication, her doctor suggested she try to eat more potassium-rich foods to see if it would bring her potassium levels up to a normal range. Two weeks later, after she enriched her meals with potatoes, avocados, and other potassium-rich foods, her potassium levels were above the normal range. The efficacy of the changes she made in only a couple of weeks is a powerful example of the effectiveness of food as medicine.

Another less personal example of food being powerful medicine can be found in the 2014 *Journal of Family Practice* study showing how plant-based foods can reverse coronary artery disease in severely ill patients in as little as three weeks. This study, led by Dr. Caldwell Esselstyn, documents a non-pharmaceutical approach that not only helps prevent heart disease but can also reverse it, even in late stages.[62]

Dr. Esselstyn's study consisted of 198 patients who adopted plant-based eating. After a mean of 3.7 years, 177 (89 percent) participants had maintained a plant-based lifestyle and were less likely to experience a significant cardiac event. In fact, only one major cardiac event (stroke) occurred in the plant-based group (a recurrent event rate of 0.6 percent), compared to thirteen major cardiac events occurring in the twenty-one people in the non-plant-based group (a recurrent event rate of 62 percent).[63]

Similarly, years earlier, the *Journal of the American Medical Association* reported findings of another clinical trial involving patients who also observed a plant-based lifestyle. This study, led by Dr. Dean Ornish, Clinical Professor of Medicine at the University of California, San Francisco, tracked, among other things, all cardiac events over five years between the control group and the experimental group that adopted a plant-based lifestyle. Over the five-year study, the non-plant-based group had 2.25 cardiac events per patient compared to 0.89 cardiac events per patient in the plant-based group.[64]

After only one year, the plant-based patients had a 91 percent reduction in reported frequency of angina (chest pain or pressure often due to insufficient blood flow to the heart) while the non-plant-based patients had a 186 percent increase in reported frequency of angina. Additionally, the non-plant-based patients experienced a progression of coronary atherosclerosis of more than twice the plant-based group.[65]

Getting back to my friend Ted, one of the medications his doctor prescribed him was a statin. Statins comprise one of the most widely prescribed medications globally, accounting for more prescriptions in the US since 1990 than any other. In the US, 28 percent of people over the age of forty take a statin.[66]

According to an analysis by the Cochrane Collaboration, of one thousand people with no history of cardiovascular disease (CVD) treated with statins for five years, only eighteen (1.8 percent) avoid a major CVD event.[67]

NNT stands for Number-Needed-to-Treat. The NNT website says,

> We are a group of physicians that have developed a framework and rating system to evaluate therapies based on their patient-important benefits and harms as well as a system to evaluate diagnostics by patient sign, symptom, lab test, or study. We only use the highest quality, evidence-based studies (frequently, but not always Cochrane Reviews), and we accept no outside funding or advertisements.[68]

NNT.com also concludes, "After 5 years of daily statin therapy study, subjects achieved a 1.2 percent lower chance of death, a 2.6 percent lower chance of heart attack, and a 0.8 percent lower chance of stroke." It's worth noting the NNT website reports the potential risks from taking statins include muscle breakdown, diabetes, and cancer.[69] I believe patients would be more motivated to try other modalities, such as lifestyle changes, if they were more aware of statins' risks and low effectiveness. No patient goes to a doctor asking for a medication that will reduce their chance of death by only 1.2 percent.

The decision becomes crystal clear when comparing statins' measly protection to only one major cardiac event (stroke) out of 177 diet-compliant patients in Dr. Esselstyn's study. His plant-based group had a 99.4 percent lower recurrence rate without any negative side effects.[70]

With 28 percent of people in the US over the age of forty taking statins[71], it's a financial windfall for big pharma and the Meatrix when people choose to ignore the health impact of the food they eat and instead rely on pills. Don't forget that, like all corporations, statin manufacturers and meat producers are not looking out for you but their own bottom line. Not only does relying on pills reinforce our dependence on pharmaceuticals, but it also robs millions of people of the opportunity to live longer, more productive lives by adopting healthier plant-based lifestyle changes.

While it seems that many people are content to take medication rather than make lifestyle changes, in the past sixty years, we have seen Americans make a gradual yet dramatic shift away from eating artery-clogging red meat to chicken (jumping from 2.61 billion tons of meat chickens in the US in 1961 to over 19.57 billion tons in 2018) as a way to lower their cholesterol and prevent heart disease.[72] Switching from red meat to chicken sounds like a smart choice, but a recent study's findings indicate that the move to chicken hasn't been an effective strategy for those wishing to lower their cholesterol levels.

In 2019, the *American Journal of Clinical Nutrition* reported that white chicken meat raised LDL cholesterol (low-density lipoprotein, i.e., "bad" cholesterol) just as much as red meat. The researchers, who assumed that only the red-meat-eating group would have increased LDL levels, were surprised by the study's findings. The study also showed that those who ate

U.S. CHICKEN CONSUMPTION, 1961–2018 (MILLIONS OF TONS/YEAR)

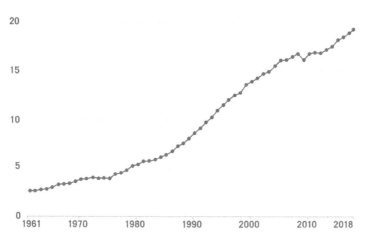

Source: UN Food and Agriculture Organization (FAO),
ourworldindata.org/meat-production

the plant-based protein diet had no increase in LDL levels.[73] According to the American Heart Association, LDL cholesterol can build up in a person's blood vessels, increasing the risk of a heart attack or stroke.[74]

While many people made the shift away from red meat to chicken as a way to lower their bad cholesterol, fewer people replaced the animal protein on their plate with plant-based alternatives. Unfortunately, this can have dire consequences for people's health. In the last sixty years, Americans have simply replaced one bad source of saturated fat with another bad source.

2. Cancer

According to further research done by the CDC, the total number of Americans dying from heart disease is rising. The CDC also reports that cancer deaths have tripled since 1950 and continue to rise as well.[75] Some medical professionals attribute the rise in cancer to Americans living longer, but "many of the risk factors for heart disease also increase the risk for cancer, especially smoking and poor diet," said Dr. Mariell Jessup of the Penn Heart and Vascular Center at the University of Pennsylvania (and former president of the American Heart Association).[76]

The same food-as-thy-medicine approach employed by Dr. Esselstyn and Dr. Ornish to reverse heart disease can also help prevent cancer. In fact, the Mayo Clinic says that cancer is the number-two cause of death in America, and diet and nutrition alone could prevent up to 33 percent of all cancers. In addition, the Mayo Clinic states, "In fact, vegans—those who don't eat any animal products including fish, dairy or eggs—appeared to have the lowest rates of cancer of any diet." They go on to say that plant-based eating is so effective at preventing cancer because plant chemicals (phytochemicals) reduce cell damage and are anti-inflammatory.[77]

In 2015, the International Agency for Research on Cancer (IARC), of the World Health Organization (WHO), classified processed meat as a Group 1 carcinogen to humans. WHO considers processed meat to be that which "has been transformed through salting, curing, fermentation, smoking, or other processes that enhance flavor or improve preservation." These meats include bacon, hotdogs (frankfurters), ham, sausages, corned beef, beef jerky, and canned meat, as well as meat-based preparations and sauces. WHO designated processed meats as a Group 1 carcinogen because "there is convincing evidence that the agent causes cancer."[78] WHO spoke specifically of colorectal cancer, but further evidence from the same IARC study also links processed meats with pancreatic and prostate cancer.[79]

Recently I was shopping at our local Good Foods Co-op. These days, to be competitive with Whole Foods, our co-op offers a wide variety of animal-based products. As I approached the checkout lane, I noticed a free-standing display rack of beef jerky—products that, as mentioned above, have been labeled as carcinogenic by WHO. The Meatrix is so pervasive and persuasive that they could garner a prime location in a health food grocery to display their cancer-causing products. (After I spoke with management, the display has been moved.)

According to 2013 data from the Global Burden of Disease Project, 644,000 deaths attributed to diets high in processed meats occur each year globally. Their analysis of data from ten different studies estimated that a daily intake of fifty grams of processed meat increases one's risk of colorectal cancer by 18 percent.[80] Two pieces of lunch meat or one hotdog exceeds this amount. My childhood was filled with hotdogs and bologna sandwiches for lunch. How about yours?

It's not only red and processed meats associated with an increased risk of cancers and mortality; dairy products are proven to increase the risk of death from prostate cancer. The *International Journal of Cancer* reported the results of a ten-year Physicians Health Study that monitored the dairy intake of 926 men diagnosed with prostate cancer. Men who consumed three or more servings of dairy products daily increased their risk for death by 76 percent. This same group had a 141 percent higher risk of death due to prostate cancer than those who consumed less than eight ounces of dairy daily. The study found no evidence of low-fat dairy products offering any protection. The researchers believe the saturated fat, hormones, and calcium found in dairy products account for the increased risk.[81]

3. COVID-19

While the Peterson-KFF chart above lists COVID-19 as the third leading cause of death in 2021, the number of daily deaths it reports, 727, was from August 1–24. As of December 29, 2021, the average number of US deaths due to COVID is 1,243, much higher than August but still a fraction of the record 4,327 set on January 12, 2021.[82]

In his frequent COVID-19 updates, Kentucky governor Andy Beshear reminds Kentuckians, "We are going to get through this, and we're going to get through this together." One of the things that he says we all need to do is be healthier. He includes himself in this. I mentioned earlier that Kentucky is not a haven of good health. Looking at obesity alone, in 2018, the University of Kentucky reported the state obesity rate was 36.6 percent (and 43.8 percent for ages 45 to 64), up from 21.7 percent in 2000. It's important to know that obesity has been linked to more than sixty chronic diseases, including higher rates of cancer[83] (see the "Obesity" section later in this chapter).

It's encouraging to see how some leaders realize the connection between diet, health, and our ability to withstand the current pandemic and possible future ones. In 2020, Mexican epidemiologist Hugo López-Gatell Ramírez, spokesperson for the task force addressing the 2020 COVID-19 pandemic in Mexico, said, "Decades of poor eating habits in the country have created an epidemic of obesity, diabetes and other related health complications that make its people more vulnerable to the novel coronavirus."[84]

While many others have also observed a link between dietary habits and poor COVID-19 outcomes, no scientific data was available until an article was published in the BMJ in June 2021.

Between July 17 and September 25, 2020, 2,884 healthcare workers from six countries (France, Germany, Italy, Spain, UK, US) with high exposure to the coronavirus completed a survey. Participants reported whether they followed any dietary pattern in the year preceding infection and their COVID-19 symptoms and duration of symptoms. The study determined plant-based eaters had a 73 percent lower chance of moderate to severe COVID-19.[85] That is excellent news for all those who have escaped the Meatrix and offers hope for those wishing to increase their chances of surviving this and perhaps future pandemics.

You would think when a prestigious medical journal releases data showing plant-based eaters have a 73 percent less chance of experiencing moderate to severe COVID-19, it would get a lot of media coverage—especially at a time when the delta variant was raging, and COVID-19 cases were rising dramatically around the globe. However, most major news media outlets declined to report this groundbreaking and relevant study.

A few months into this pandemic, healthcare workers saw a pattern in those with the worst COVID-19 outcomes. One article noted five preexisting conditions and their role in COVID-19 outcomes: heart disease, chronic respiratory disease, diabetes, depression, and anxiety.[86] All five of those conditions can be avoided or improved with a plant-based lifestyle.

We just discussed heart disease and cancer (and will discuss diabetes later in this chapter), but what about depression and anxiety? Can dietary patterns lessen the impact of these conditions? The short answer is yes! Escaping the Meatrix gives you more than a healthy body that's able to meet the demands of a busy lifestyle. It's also beneficial for your mood. I've already mentioned some of the ways that plant-based eating boosts my physical and emotional states. But what does science say?

A 2017 *Psychiatry Research* article found that "a dietary pattern characterized by a high consumption of red and/or processed meat, refined grains, sweets, high-fat dairy products, butter, potatoes, and high-fat gravy, and low intakes of fruits and vegetables is associated with an increased risk of depression."[87]

A 2016 study published in *Nutritional Neuroscience* "showed that overall, vegans, and to a lesser extent vegetarians, reported less stress and anxiety than omnivores. More specifically, male vegans and vegetarians reported significantly lower anxiety scores than did male omnivores, and female vegans reported significantly lower stress scores than did female omnivores."[88] Another article from two years earlier reported similar findings.[89]

So if you want to lower your anxiety and stress, here are ten vegan foods to help you do just that: almonds, ginger root, leafy greens, beans, bananas, soybeans, black currants, guava, bell peppers, and oats.[90]

4. Accidents

A plant-based lifestyle offers no protection from accidental deaths from vehicular accidents, falls, etc.

5. Stroke

If a plant-based lifestyle can reverse even late-stage cardiovascular disease, what about 2021's fifth leading cause of death in the US, stroke?

In March 2021, the American Academy of Neurology published the findings of a study to determine if a plant-based lifestyle offered any protection from total, ischemic, and hemorrhagic stroke. Participants in the study consisted of 73,890 women in the Nurses' Health Study (1984–2016), 92,352 women in Nurses' Health Study II (1991–2017), and 43,266 men in the Health Professionals Follow-Up Study (1986–2012). All participants were cancer-free and without cardiovascular disease at the beginning of the study. After following more than 200,000 participants, the study concluded a "lower risk of total stroke was observed by those who adhered to a healthful plant-based diet."[91]

6. Chronic Respiratory Disease

In March of 2015, the National Institute of Health found that "the dietary patterns associated with benefits in respiratory diseases include high fruit and vegetable intake...while fast food intake and Westernized dietary patterns have adverse associations."[92]

Another NIH article from June of 2019 reported an association between Western diets consisting of processed and animal-based foods and decreased lung function and chronic obstructive pulmonary disease (COPD). The article further stated plant-based foods and healthy fats have the opposite effect by preserving lung function and preventing COPD. In conclusion, the report found that "the magnitude of effect of diet on lung function is estimated to be comparable to that of chronic smoking."[93] Is it a coincidence that the same plant-based lifestyle that reverses heart disease, reduces cancer rates, decreases the risk of moderate to severe COVID, and lowers incidents of stroke also preserves lung function? Clearly a pattern is emerging!

7. Alzheimer's Disease

According to the Physicians Committee for Responsible Medicine, a plant-based lifestyle can also improve cognitive brain function and offer protection from Alzheimer's disease.[94]

Almost everyone wants to live a long, happy life and gracefully age with minimal decline physically and mentally. Do diets rich in the saturated fat and cholesterol found in animal-based foods affect our brains and cognitive function? The answer is a definitive yes.

Currently, 44 million people worldwide live with Alzheimer's disease, with only one diagnosis in every four people living with Alzheimer's disease. In addition, 5.5 million Americans of all ages have Alzheimer's disease, and two-thirds of them are women.[95] In the US, one person is diagnosed with Alzheimer's every sixty-six seconds. However, it's estimated the time will be halved to thirty-three seconds by the middle of the century!

Before COVID, Alzheimer's disease was the sixth leading cause of death in the US, ahead of breast and prostate cancers. In addition, the average life expectancy of someone diagnosed with Alzheimer's is four to eight years.[96] So you can see there's an Alzheimer's disease pandemic occurring in the US (and other Western countries).

According to an article from the National Institute on Aging:

> The disease is named after Dr. Alois Alzheimer. In 1906, Dr. Alzheimer noticed changes in the brain tissue of a woman who had died of an unusual mental illness. Her symptoms included memory loss, language problems, and unpredictable behavior. After she died, he examined her brain and found many abnormal clumps (now called amyloid plaques) and tangled bundles of fibers (now called neurofibrillary, or tau, tangles).
>
> These plaques and tangles in the brain are still considered some of the main features of Alzheimer's disease. Another feature is the loss of connections between nerve cells (neurons) in the brain. Neurons transmit messages between different parts of the brain, and from the brain to muscles and organs in the body. Many other complex brain changes are thought to play a role in Alzheimer's, too.[97]

But is Alzheimer's disease caused by genetics or something else? Generally speaking, there are two types of Alzheimer's disease: early onset and late onset. Early-onset Alzheimer's is genetic, while the other, late onset, may have a genetic component. Still, risk factors for late-onset Alzheimer's include hypertension and diabetes, two conditions plant-based living helps prevent. Fortunately, only 6 to 7 percent of Alzheimer's disease is early onset, meaning most people can avoid Alzheimer's entirely by reducing and managing risk factors for this disease, which currently afflicts one out of nine Americans sixty-five years and older.[98]

According to the article titled "Ethnic Differences in Dementia and Alzheimer's Disease," by Jennifer J. Manly and Richard Mayeux, Japanese-Americans' risk of Alzheimer's disease is closer to other Americans than Japanese living in Japan. The same article reports that the Alzheimer's disease rate is five times lower for Africans living in Nigeria than for African Americans living in Minneapolis. When people immigrate to the US from countries with more plant-based diets, they often replace those healthy foods with the standard American diet. The increased risks for Japanese Americans and African Americans living in the US who adopt a Meatrix-based diet suggest the risk for most Alzheimer's is not genetic.[99] For example, an article from the journal *Neuroepidemiology* found meat eaters (including poultry and fish) had dementia rates more than double their vegetarian counterparts. The gap increased when past meat consumption was taken into account.[100]

According to the Alzheimer's Research & Prevention Foundation website, avoiding a diet high in trans and saturated fats is an excellent way to feed your brain for better memory.[101]

The great news is that avoiding these fats is easy once you escape the Meatrix because very few plant-based foods contain saturated fats and zero whole plant-based foods contain trans fats. Fats from animal products like beef, cheese, and ice cream, increase inflammation and free radicals in our bodies. Of course, free radicals are a normal part of the body's metabolism, but high quantities of them can damage and even kill brain cells, according to the Alzheimer's Research & Prevention Foundation, so it's best to avoid animal-based products, especially red meat.

Of course, we need healthy fats in our diets. According to Harvard Medical School, healthy fats are liquid, not solid, at room temperature and consist of monounsaturated and polyunsaturated. A few of the benefits of healthy fats are reducing our risk of stroke and heart disease, lowering bad LDL cholesterol and raising good HDL (high-density lipoprotein) cholesterol, and helping our bodies absorb fat-soluble vitamins A, D, E, and K, as well as offering protection from memory loss and dementia. Plant-based foods high in healthy fats include nuts, seeds, edamame, tofu, cacao nibs, and olives.[102]

Research shows that those eating a typical Western diet consuming high levels of cholesterol, saturated fats, and excess calories while eating low amounts of fiber, fruits, and vegetables have an elevated risk for Alzheimer's.[103] We know that cholesterol and saturated fats in the body create a plaque that can clog arteries. But the identical plaques that clog arteries can also restrict blood flow to areas of the brain, reducing neurotransmitters—brain chemicals that transmit our brain's messages from one neuron to another.[104]

Cholesterol also leads to increased oxidation and inflammation (see the section on the Maillard Reaction in Chapter 9 for more information), which harms healthy brain function. The great news is you can lower your risk factor for Alzheimer's disease by avoiding the saturated fats and cholesterol found in animal-based foods like red meat, chicken, fish, eggs, dairy products, and butter. You can then replace those foods with plant-based foods high in antioxidants like blueberries, pecans, strawberries, artichokes, goji berries, raspberries, kale, red cabbage, and beans. Antioxidants, especially vitamins C and E, lower oxidative stress in our bodies by removing free radicals that are byproducts of metabolism.[105]

Elevated levels of the amino acid homocysteine are risk factors for both heart disease and memory loss, and we get homocysteine mostly from eating meat.[106] In fact, homocysteine is formed during the metabolism of the amino acid methionine, which is abundantly found in meat, fish, and dairy products.[107]

Healthy people convert homocysteine into a benign product, but homocysteine builds up in the body when it's not metabolized properly. This elevation is significant because high homocysteine levels are one biomarker for increased risk of dementia, Alzheimer's disease, and heart disease.[108] You can lower your homocysteine levels simply by escaping the Meatrix.

I talk about advanced glycation end products (AGEs) in Chapter 9, but I want to mention here that they are a risk factor for developing dementia and Alzheimer's disease. A Meatrix-centered diet is high in AGEs, and elevated concentrations of AGEs may predict cognitive decline. In a 2016 article, Dr. Michael Greger wrote, "If you measure the urine levels of glycotoxins flowing through the bodies of older adults, those with the highest levels went on to suffer the greatest cognitive decline over the subsequent nine years."[109] Thus, lowering your AGEs is another excellent reason to escape the Meatrix.

If you have a family member with Alzheimer's disease, you may have a genetic predisposition for it, but that doesn't mean you'll contract it. Fortunately, you can decrease your likelihood of Alzheimer's by reducing the risk factors mentioned above, exercising, and increasing your consumption of whole plant-based foods. In fact, Dr. Chris Walling, Vice President for Education at the Alzheimer's Research & Prevention Foundation, says, "Plant-based foods are beneficial to the brain and may help prevent Alzheimer's disease and other forms of dementia.

Research has also shown how plant-based diets significantly reduce depression, anxiety, and fatigue."[110]

8. Diabetes

As I mentioned in the Introduction, Kentucky has, to put it nicely, a lot of health challenges. For example, the CDC ranks Kentucky fifth among the fifty states with the most cases of diagnosed diabetes.[111]

But how prevalent is diabetes in the rest of America? According to the CDC, one-third (88 million) of all adults in the US are pre-diabetic (where blood sugar levels are abnormally elevated), and 34.2 million Americans (more than 10 percent of the total population) are living with diabetes.[112]

Sadly, as a result, diabetes is the seventh leading cause of death in Kentucky (although in 2021 it dropped to eighth due to deaths from COVID-19).[113] But the good news is that most of us can avoid this horrifying disease that causes blindness and ravages the body until it brings death. The National Institute of Health reported that "diet and lifestyle, particularly plant-based diets, are effective tools for type 2 diabetes prevention and management."[114]

9. Other Diseases of the Respiratory System

Chronic respiratory disease, chronic lower respiratory disease, and other diseases of the respiratory system such as asthma are also improved through a plant-based lifestyle.

A randomized controlled trial of the antioxidant intake of adults with asthma found that those who ate high amounts of fruit and vegetables for three months experienced a reduction in asthma exacerbation compared to test subjects who consumed low amounts of fruits and veggies.[115]

But it's not only adults with asthma who benefit from the abundance of antioxidants in fruits and vegetables. For example, the NIH reported that school children aged eight to twelve who ate fruit saw a reduction in wheezing and that children who ate green vegetables had a lower prevalence of both wheezing and asthma.[116]

That article further reported that a meta-analysis of twelve cohort studies, four population-based case-control studies, and twenty-six cross-sectional studies also revealed increased fruit and vegetable consumption reduces wheezing in children and is associated with lower asthma risk in children and adults.[117]

In conclusion, the article stated:

> A whole foods approach to nutrient supplementation—for example, increasing intake of fruit and vegetables, has the benefit of increasing intake of multiple nutrients, including vitamin C, vitamin E, carotenoids and flavonoids and shows more promise in respiratory diseases in terms of reducing risk of COPD and incidence of asthma exacerbations.[118]

A sixteen-week study reported by Cambridge University Press in June of 2010 found that the immune response to Pneumovax II (a pneumococcal polysaccharide vaccine) was more significant in elderly patients who consumed at least five servings of fruits and vegetables daily. The study concluded that "increased fruit and vegetable intake may improve antibody response to vaccination in older people, linking an achievable dietary goal with a potential improvement in immune function."[119]

10. Renal Disease (Chronic Kidney Disease)

In the 1960s, scientists determined that overall lower protein consumption protected kidneys in patients with chronic kidney disease (CKD). However, the source of the protein was rarely studied in those patients, but multiple 2019 studies showed there was a big difference. For instance, one study showed that patients with CKD could eat increased amounts of plant-based protein while still protecting their kidneys. In addition, the article found patients with CKD could easily meet all their nutritional-related goals with plant-based foods.[120]

A second article reported that the primary health conditions that cause CKD are lower in those who eat a plant-based diet. Two of the three leading causes of CKD are hypertension (high blood pressure) and type 2 diabetes, which a plant-based lifestyle helps reverse or prevent. The article continues by noting that a plant-based "dietary pattern is associated with a reduced risk of all-cause mortality in CKD patients" and that "plant-based diets should be included as part of the clinical recommendations for both the prevention and management of CKD."[121]

The National Kidney Foundation published an article that found "other disciplines of the health care field have used plant-based diets to their benefit in treating heart disease, diabetes, and obesity." In addition, it stated that "food can be seen as being complementary to pharmacologic therapies for patients with CKD. Instead of running away from these foods, and perhaps incurring harm by doing so, we should be embracing these foods to our collective benefit."[122]

11. Suicide

Suicide, the eleventh leading cause of US deaths in 2021, is never an easy thing to discuss. There are innumerable reasons one might choose this route, but poor health, despair about the world, and seeing so much cruelty to humans and animals can play a role. If you suffer from severe depression or have thoughts of self-harm or know someone who does, I strongly encourage you to seek help. The toll-free number for the National Suicide Prevention Lifeline in the US is 1-800-273-8255. Also, please look at section three (COVID-19), which discusses lowered rates of depression and anxiety in those following a plant-based lifestyle.

Other Illnesses

The following health conditions did not appear on the list of the top eleven causes of US deaths in 2021 but are worth mentioning because they affect the lives of millions of Americans. A plant-based lifestyle is a way to reverse or avoid these conditions.

Hypertension

High blood pressure (hypertension) is commonly referred to as "the silent killer" and is associated with a Western diet high in processed and animal-based foods. As already discussed, the Meatrix contributes to many of the chronic health conditions in the US, and cholesterol-reducing statins aren't the only medication big pharma has created to address a common ailment that plant-based eating can reverse or even prevent. In August 2020, the GoodRx.com website listed the statin Lipitor as the number one prescribed pharmaceutical in the US, and the second medication on their list was lisinopril, often prescribed for high blood pressure.[123] Lisinopril is an angiotensin-converting enzyme (ACE) inhibitor. The renin-angiotensin system (RAS) plays a critical role in regulating our body's blood pressure. ACE inhibitor drugs like lisinopril prevent the production of angiotensin II, a vasoconstrictor that narrows blood vessels. This narrowing of blood vessels can lead to high blood pressure, which, among other things, forces your heart to work harder.[124]

In 2019, lisinopril was prescribed over ninety-one million times in the US alone.[125] And like those taking statins, patients taking ACE inhibitors for high blood pressure, heart failure, and diabetic kidney disease are on them long term, often for the rest of their lives. Also, like statins, ACE inhibitors have side effects including headache, cough, muscle weakness, connective tissue problems, erectile dysfunction, loss of libido, hair loss, and liver and pancreas inflammation.[126]

In January 2022, a pandemic-related increase in hypertension in the US was reported. An observational study of about a half-million Americans enrolled in employee-sponsored wellness programs found no elevation in hypertension in the 15 months prior to the COVID-19 pandemic (January 2019 through March 2020).[127] However, the same study reported that hypertension increased in the remaining nine months of 2020, a period of lockdowns and self-isolation. From April 2020 to December 2020, the average systolic, or top number, rose by an average of two millimeters of mercury (mm Hg) while the diastolic, or bottom number, saw a slight increase, too. "That's concerning," said Luke Laffin, M.D., the study's lead author and co-director of the Center for Blood Pressure Disorders at the Cleveland Clinic. "On the individual level, that [2 mm Hg increase] doesn't seem like a lot. But we know that a small increase in blood pressure can cause a significant increase in strokes and heart attacks across the population."[128] The same study reported that nearly half of all adults in the US have hypertension, but only 24 percent have it under control.

Fortunately, lifestyle changes, such as plant-based eating, can often control high blood pressure—an important fact because an article in the *Journal of Geriatric Cardiology* reported that eighty million Americans twenty years of age and older were living with high blood pressure. The article states that first-line therapies are weight loss and exercise. However, the article continued that a small cross-sectional study found that "a plant-based diet is the more important intervention." That study found that sedentary plant-based eaters had lower blood pressure than those "consuming a Western diet and running an average of 48 miles per week." After citing a variety of clinical trials and studies, the report concluded, "The totality of evidence taken from these studies indicates that plant-based diets have a meaningful effect on both prevention and treatment of hypertension."[129]

Obesity

Like the other diagnoses mentioned above, obesity has been linked to a poor diet. Obesity is a significant problem in the US and other parts of the world and has been climbing steadily for years. Obesity in the US was just over 10 percent in 1975 and has risen every year since to 42.4 percent in 2020. More than sixty chronic diseases are associated with obesity.[130]

In 2012, the American Cancer Society reported that one-third of the 572,000 cancer deaths each year are linked to excess body weight, poor nutrition, and physical inactivity.[131] The World Health Organization also stated that "the risk of coronary heart disease, ischaemic stroke and type 2 diabetes grows steadily with increasing body mass, as do the risks of cancers of the breast, colon, prostate, and other organs."[132] Obesity costs anywhere from $147 billion to $210 billion per year in the US alone and is also connected to absenteeism, with overweight and obese people missing 56 percent more workdays than employees of average weight, which costs employers $4.3 billion annually due to lower work productivity.[133]

Sugary sodas were linked to obesity in the 1990s, launching an effort to reduce their consumption. According to the Associated Press, soda drinking peaked in 1998, with the average American consuming about fifty-four gallons per year.[134] In 2017 soda consumption in the US was at a thirty-one-year low, a trend that is expected to continue.[135] That's great news, but if

SHARE OF US ADULTS WHO ARE OBESE, 1975–2016

Obesity is defined as having a body mass index (BMI) ≥ 30. BMI is a person's weight in kilograms divided by their height in meters squared.

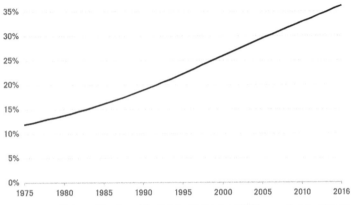

Source: Hannah Ritchie and Max Roser, "Obesity," *Our World in Data* (2017), ourworldindata.org/obesity

sugary sodas were causing obesity, why haven't we seen a drop in the obesity rates, too?

Obesity is clearly on the rise, but how is obesity determined? Most doctors in the US use the body mass index (BMI) to assess whether someone is a healthy weight, overweight, or obese. This system has its advantages and disadvantages. Critics highlight that BMI doesn't consider a patient's age, sex, race, or body composition. However, doctors argue that using BMI gives them an objective, clinical, and non-observational way to discuss body weight, which helps preserve the patient–clinician relationship. BMI is calculated by dividing weight (in pounds) by height (in inches) twice and then multiplying that number by 703.[136] A healthy weight is a BMI between 20 and 25, overweight is a body mass index from 25 to 30, and obese is anything over 30.

If soda consumption is at a thirty-year low and obesity rates are at an all-time high, could the actual culprit be the Meatrix? Between 2002 and 2007, researchers studied 71,751 Seventh Day

Adventist men and women with an average age of fifty-nine. The study found that vegans had a lower body mass index (BMI) than any other group—carnivore, omnivore, pescatarian (eating fish), or lacto-ovo-vegetarian (eating dairy and eggs).[137]

So it seems in addition to lowering your risks of death from nine of the ten leading causes of US deaths in 2021, escaping the Meatrix is excellent for your waistline! (In fact, it's worth mentioning that in the diet-adherent group in Dr. Esselstyn's study mentioned above, each participant lost an average of 18.7 pounds.)[138]

As I said in the opening to this chapter, the Meatrix doesn't want you to know that not only your health but the health of the planet is what's currently at stake. Instead, it gaslights you with bogus, industry-funded studies, sacrifices your health, and denies its role in climate change (see Chapter 4) entirely for corporate profit. Fortunately, the National Institute of Health knows and has passed along to the medical community what *would* help bring healing to their patients: "Physicians should consider recommending a plant-based diet to all their patients, especially those with high blood pressure, diabetes, cardiovascular disease, or obesity."[139]

This update to physicians was in 2013! Why isn't this information displayed in every doctor's office rather than advertisements for insulin and cardiac medications? Why aren't patients with cardiovascular disease, cancer, respiratory disease, chronic obstructive pulmonary disease (COPD), early-onset Alzheimer's disease and cognitive decline, hypertension, obesity, and diabetes told that one simple yet vital lifestyle change could prevent, reverse, or stabilize their chronic health conditions?

Study after study and report after report recommend a plant-based lifestyle full of fruits, vegetables, whole grains, legumes, seeds, and nuts in order to to have a healthier life. And we now know that a plant-based lifestyle can help you not only feel great but avoid common chronic diseases, many of which were the top killers in the US in 2021.

FASCINATING FACTS

- The average person consumes 80,000 pounds of food in a lifetime.
- The average medical student spends only 19.6 hours studying nutrition in medical school.
- 33 percent of cancer deaths are related to poor nutrition and diet.
- An estimated 644,000 people globally die due to processed meat per year.
- Plant-based eaters have a reduced risk of severe to moderate COVID-19 of 73 percent.
- The dementia rate for meat eaters is two times that of vegetarians.
- A study following 71,751 Seventh Day Adventists men and women determined that plant-based eaters had the lowest body mass index of any other dietary pattern.

PANDEMICS AND EMERGING INFECTIOUS DISEASES

"Those who cannot remember the past are condemned to repeat it." Nineteenth-century Spanish philosopher George Santayana is credited with that aphorism. It's true our knowledge of history provides us vital context for the present day and can help guide our decisions as we plan for our future. With that in mind, this section contains many historical references that offer context for today's coronavirus pandemic. With knowledge comes power, and I hope this information empowers you to make positive decisions for your future, a future firmly grounded in fully understanding and learning from our past mistakes, lest we end up repeating them.

As I'm writing this, the global coronavirus pandemic is far from over. In the early days of the pandemic, some downplayed the seriousness of the virus, claiming it was no worse than the flu. But the pandemic's impact has been far worse than the seasonal flu. We've seen multiple surges as well as new, more contagious variants emerge. One metric we can use to measure the seriousness of the pandemic is excess mortality rates and decreased life expectancy. US mortality rates for the first twelve months of the pandemic increased by 646,514, with 83.4 percent attributed directly to COVID-19.[140] Additionally, the coronavirus has become the third leading cause of US deaths in 2021, and the US has experienced the most significant drop in life expectancy since World War II!

A *USA Today* story from July 2021 attributes only 74 percent of decrease in life expectancy to COVID-19, with drug overdoses (perhaps indirectly related to COVID-19) accounting for the rest.[141] According to the CDC, in 2020, the life expectancy of white Americans went from 78.8 years to 77.3.[142] Sadly, the falling life expectancy rate is two and three times higher for the Black and Hispanic Americans, respectively.

In October 2021, ABC News reported that the US had experienced even more deaths in 2021 from COVID-19 than the ten months of the pandemic in 2020.[143] This statistic is especially startling considering 2021 saw the widespread availability of effective vaccines and a greater understanding of effective treatments for patients with COVID-19.

Climate change and biodiversity loss (both of which, as we'll discuss, are Meatrix-fueled) are often drivers of emerging infectious diseases (EID), including zoonosis and pandemics.[144]

Most scientists agree the environments that give rise to both emerging infectious diseases and pandemics are fueled in many ways, if not entirely created, by human activities. However, the good news is since they're often of our own making, it's possible to mitigate the worst-case scenarios by changing how we live our lives. I will speak about how the Meatrix contributes to climate change in the next chapter, but how does the Meatrix contribute to emerging infectious diseases?

The novel coronavirus is the most recent emerging infectious disease to have resulted in a worldwide pandemic. While there has been much debate concerning the origins of the coronavirus, in February 2022, *The New York Times* reported two extensive studies that point to the wild animal market in Wuhan as the origin of COVID-19.[145]

However, some continue to believe the virus originally escaped the Wuhan Institute of Virology, located only eight miles away from the wild animal market.[146] This scenario was posed most famously by some senior Trump administration officials in the spring of 2020. Scientists who disagree with the leaked lab theory believe the novel coronavirus has a natural origin, that the virus made the leap from animal to human via contact at the Wuhan market, known as a "wet" market, where one can purchase live rare wild and exotic animals considered delicacies. These types of markets are relatively unknown in the West but are popular in some parts of Asia, Africa, and the Middle East, where traditional medicine claims that consuming certain exotic species imparts unique healing properties to the human body.

Sadly, human culpability plays a role in both of the above scenarios when it comes to the coronavirus. In the first, our negligence allowed the virus to escape a lab, and in the second, our obsession with meat and wild animals brought us into contact with a dead-end host of the virus.

It's been well established that coronaviruses *have* made the leap from animals to humans in the past, such as the 2002 SARS-CoV outbreak, the first pandemic of unknown etiology in the twenty-first century (I'll go into a bit more detail on that outbreak below). A February 2021 article reported a probable bat origin to the SARS-CoV-2 virus, saying, "SARS-CoV-2 related viral genome sequences from bats have been reported from Eastern China and Japan, and from pangolins in China."[147] But, regardless of its origins, SARS-CoV-2 (the virus that causes COVID-19), as of spring 2022, is still wreaking havoc all over the planet, and it's not over yet.

I recently ran across a 2008 video by Dr. Michael Greger, who was then Director of Public Health and Animal Agriculture for the Humane Society of the United States. In his video, Dr. Greger outlined a history of pandemics. "Medical anthropologists have identified three major periods of disease since the beginning of human evolution, and the first started just ten thousand years ago, with the domestication of animals."[148]

Certainly, hunter-gatherers had infectious diseases. But the concentration of populations due to the advent of agriculture and the domestication of animals was the perfect storm for pandemics to arise. In the video, Dr. Greger said, "When we brought animals into the barnyard, they brought their diseases with them."[149]

The WHO defines zoonosis as "an infectious disease that has jumped from a non-human animal to humans. Zoonotic pathogens may be bacterial, viral or parasitic, or may involve unconventional agents and can spread to humans through direct contact or food, water or the environment."[150] Zoonotic viruses, such as the coronavirus, are considered novel because humans have no prior history with them and therefore no acquired immunity. Unfortunately, we've seen firsthand just how disruptive and deadly emerging infectious diseases can be.

But the novel coronavirus is simply the most recent pandemic to emerge out of a line of emerging infectious diseases that have unleashed ineffable devastation on humans throughout the centuries. So what are some of these emerging infectious diseases, how have they impacted humanity, and how are they connected to the Meatrix?

Smallpox

In the twentieth century, smallpox was eradicated, but not before it wreaked global havoc. Caused by the variola virus (VARV), it is a highly contagious and lethal disease thought to have first infected humans in East Africa around three thousand to four thousand years ago. According to the National Institute of Health, genome testing demonstrates that VARV emerged from camelpox (CMLV) around six thousand years ago.[151] A BBC article estimated that three hundred million people died of smallpox in the twentieth century alone. The total number of victims of this gruesome disease is estimated to be five hundred million people. Researchers believe the combination of domesticating animals (including camels) and large human settlements made possible by agriculture created the favorable conditions for smallpox to emerge.[152]

Measles

The measles virus (MeV) is most closely related to the Rinderpest virus (RPV). According to the *Virology Journal*, it's generally accepted that the MeV evolved in an environment where cattle and humans lived close to each other.[153] The domestication of cows, like camels and smallpox, allowed for the Rinderpest virus to more easily mutate and infect humans with lethal consequences. In his video, Dr. Greger claimed that in the last 150 years, 200 million people have died from measles.[154] According to the WHO, over 140,000 people died from measles in 2018 alone, with most of them children under the age of five.[155] These deaths occur despite the availability of a safe and effective vaccine.

I learned in grade school that when the Spanish landed in the New World in 1492, they brought diseases that the native populations had never encountered. These zoonotic diseases brought by the Europeans decimated the native peoples. According to an article by Michael S. Rosenwald, smallpox alone killed 80 to 95 percent of the Indigenous population in the century and a half following Columbus's landing. Before then, the Americas had no flu, smallpox, or measles.[156]

But why did the Europeans remain comparatively healthy and not contract diseases from the native populations? According to a PBS article, the answer lies in the exploitation of and proximity to animals. The indigenous people of the Americas had no domesticated animals. The one exception, the llama, was geographically isolated from other regions. Also, unlike the Europeans, the Indigenous populations in the Americas didn't share their living spaces with their llamas or drink their milk, and they rarely, if ever, ate their meat.[157] The colonists conquered the indigenous cultures of the Americas, Australia, New Zealand, and South Africa more quickly due to the zoonotic diseases the Europeans inflicted upon them.

While the zoonotic diseases the Europeans carried to the New World devastated the indigenous populations, evidence suggests that explorers brought back to Europe more than potatoes, tomatoes, and tobacco. In what is sometimes euphemistically referred to as "The Great Exchange," recent scientific evidence indicates that Columbus also brought home a much stronger strain of syphilis. A 2019 *Washington Post* article reported that historians believe that syphilis was already present in Europe. In 2004, the Organization of American Historians wrote, "The strain that resulted from sexual contact between Europeans and Native Americans, however, was much stronger than the non-venereal version that a few isolated European regions had experienced. This new version was carried back to Europe and spread among the population there."[158] More importantly for our discussion, according to an October 2021 article, humans are the only host for the bacteria *Treponema pallidum*, the source of syphilis, and that no nonhuman animal reservoirs of the disease exist.[159] So it turns out that the one disease indigenous people passed on to Europeans does not have an animal origin.

Human Immunodeficiency Virus (HIV)

Another widely known infectious disease that emerged in my lifetime, before the novel coronavirus, is HIV, which, according to the WHO, has claimed the lives of over thirty-six million people worldwide as of March 2022. In 2020, 37.7 million people were living with HIV or AIDS (acquired immunodeficiency syndrome), and 680,000 of them died of HIV-related illnesses.[160] Of course, unprotected sex, a tainted blood supply, and IV drug use amplified the spread of AIDS, but what were its origins?

Scientists traced the simian immunodeficiency virus (SIV) to a specific type of chimpanzee found in the Democratic Republic of Congo (called Zaire at the time). Scientists believe HIV made the leap from chimps to humans sometime in the 1920s due to hunting and butchering infected primates.

Ebola

Ebola Virus Disease (EVD), one of the deadliest known viral diseases, was first discovered in 1976 when two outbreaks occurred in Central Africa, with the first occurring near the Ebola River. According to the Centers for Disease Control, the Ebola virus infects both human and nonhuman primates.[161]

Our first exposure to this hemorrhagic disease resulted from international logging activities that took humans deeper and deeper into the African rainforest, where they sustained themselves by consuming, among other things, bush meat. Through the actions of hunting, killing, and eating monkeys, gorillas, and chimpanzees, humans unwittingly exposed themselves to the tainted blood and excretions of primates infected with "viruses particularly fine-tuned to our primate physiology."[162]

However, scientists have determined that these primates are "dead-end" hosts for Ebola and that African fruit bats are the likely source of this deadly virus. Bats harbor up to sixty-one zoonotic viruses but they do not become ill from them because they can "maintain just enough defenses against illness without triggering the immune system from going into overdrive. In humans and other mammals, an immune-based over-response to one of these and other pathogenic viruses can trigger severe illness."[163]

An example of this can be seen in severe coronavirus cases where the immune system floods the body with pro-inflammatory cytokines. This flooding creates what's known as a cytokine storm, which leads to a poorer prognosis and higher mortality. While bats have evolved natural protection from their unusually high microbial loads, the results are often deadly when these viruses mutate and infect a new host that doesn't have these evolutionary safeguards.[164]

Severe Acute Respiratory Syndrome (SARS)

As previously mentioned, according to the NIH, SARS was the first pandemic of unknown etiology of the twenty-first century. Like COVID-19, SARS is also the result of a coronavirus (named for the "corona" or "crown" of proteins found on the virus's surface).

On November 16, 2002, the SARS outbreak began near Hong Kong in China's Guangdong province. Between November 2002 and July 31, 2003, SARS infected over eight thousand people, killing 774—an alarming kill rate of 9.6 percent.[165]

How did humans come into contact with the SARS-CoV? According to the CDC, the best available evidence points to the wild palm civet—a nocturnal mammal native to tropical Asia and Africa.[166] Civets garnered notoriety in the West via the 2007 film *Bucket List*. In the movie, billionaire Edward Cole (Jack Nicholson) and Carter Chambers (Morgan Freeman) attempt to experience everything on their bucket lists before each dies of a terminal illness. One of Edward Cole's list items is to drink Kopi Luwak, which some consider the world's best coffee. Cups of Kopi Luwak can cost anywhere from thirty-five to one hundred dollars each. Currently, you can purchase a sixteen-ounce bag of wild-gathered Kopi Luwak coffee beans for $399.99 (with free shipping) from Amazon.

But what does Kopi Luwak have to do with palm civets? Civets consume fresh, sweet coffee cherries. The coffee beans are harvested and made into coffee once they have passed through the palm civet's digestive tract. Yes, we're talking about making coffee from beans found in palm civet poop.

Before gaining popularity in the West, all Kopi Luwak coffee beans were harvested in the wild. Today almost all the harvesting is from civets held in captivity. On palm civet farms, these nocturnal, catlike creatures are contained in small cages and force-fed a diet of coffee cherries.

In 2015, twelve years after the deadly 2002 SARS outbreak in Guangdong, ABC News interviewed a civet farmer in Bali, Indonesia, who boasted she had over 102 civets in captivity. Some had been in cages continuously for more than six years.[167] So why are operations like hers allowed to continue when a highly contagious virus has been linked to civets? The answer is the Meatrix is ready to capitalize wherever and whenever it can. If there's money to be made off the flesh or secretions—or in this case the poop—of animals, humans will, time and time again, find a way to exploit animals.

Swine Flu (H1N1)

In 2009, H1N1, the swine flu, was responsible for at least seventeen thousand deaths. For the first time, using state-of-the-art genetic analysis, scientists were able to determine the precise molecular transformations that allowed the virus to jump from pigs to humans. In addition, the researchers traced the source of the virus to pigs in the Veracruz region of Mexico.[168]

In his book *Meatonomics*, author David Robison Simon describes the 2009 H1N1 outbreak this way:

> A deadly strain of swine flu raced across North America in the spring of 2009, infecting one in every five Americans and hospitalizing a quarter of a million people. ...[A] determined group of animal food producers and government officials sprang into action to address the crisis. But it wasn't the type of emergency response coalition you might expect: its focus was on saving profits, not people.[169]

A member of the Pork Producers Council feared if people kept referring to the pandemic as the swine flu, the entire pork industry would suffer. During a press conference in April 2009, USDA Secretary Tom Vilsak announced that the "hog industry is sound and safe...this really isn't swine flu. It's H1N1 virus."

I vividly remember the about-face. Almost overnight, all media, governmental, and medical institutions began calling the virus H1N1.

So which moniker is correct? They both are. Despite industry claims, a 2009 *Nature* magazine article stated that evidence conclusively determined that the virus began "in swine and that the initial transmission to humans occurred several months before recognition of the outbreak."[170]

By June, however, the name H1N1 had stuck. Why is a name important? Because the Meatrix doesn't want the name of a disease implicating it in a deadly pandemic when it's clear that it's our ten-thousand-year-old obsession with domesticating, exploiting, and eating animals that has unleashed untold death and misery upon us.

The Spanish Flu Pandemic of 1918

Another, vastly more deadly H1N1 pandemic was the Spanish flu pandemic of 1918, which killed fifty to one hundred million people—3 to 5 percent of the world's population. In the late 1990s, the US Armed Forces Institute of Pathology isolated RNA fragments from the bodies of American soldiers who had died in the 1918 pandemic. According to the NIH, "further sequencing analyses suggested that the 1918 virus may be of avian origin and transmitted from birds to humans directly or indirectly, although this remains controversial."[171]

According to the *Current Opinion in Environmental Science & Health*, zoonotic pathogens are responsible for 60.3 percent of all emerging infectious diseases. 71.8 percent of those pathogens have a wildlife origin. In wet markets, wild animal pathogens can come in contact with domesticated animals and find new hosts. These "dead-end" hosts sometimes pass the pathogen on to humans with lethal results.[172]

COVID-19

One such scenario may have given rise to the current coronavirus pandemic. In response, the Chinese government closed wet markets. But a February 2020 article in *The Guardian* reported only weeks earlier that China's State Forestry and Grassland Administration was promoting the idea of wildlife farming to its rural citizens, including civets (believed to be the dead-end host of the 2002 SARS Co-V outbreak) and pangolins (which have been found to harbor SARS CoV-2). Before the COVID-19 outbreak, we knew little about the scale of these wildlife operations, but the Chinese government has since closed over nineteen thousand wildlife farms.[173]

We read earlier that palm civets are exploited to make Kopi Luwak coffee, but what about pangolins? Why are they for sale in wet markets in China? According to *National Geographic*, the pangolin—a small creature resembling a scaled anteater—is the most trafficked non-human animal in the world. A pangolin's scales are composed of keratin—the same type of protein that makes up rhino horns, human hair, and fingernails. Like all three, keratin has no known medicinal value according to Western medicine. Regardless, traditional Chinese medicine highly praises pangolin scales, which are commonly believed to improve blood circulation, help alleviate arthritis, and help lactating women secrete milk. Unfortunately, these beliefs are driving the pangolin into extinction and increasing our risks of even more novel emerging infectious diseases.[174]

Elizabeth Maruma Mrema, the acting executive secretary of the UN Convention on Biological Diversity, hopes that the post-coronavirus world will put a greater emphasis on preserving biodiversity:

> Biodiversity loss is becoming a big driver in the emergence of some of these viruses. Large-scale deforestation, habitat degradation and fragmentation, agriculture intensification, our food system, trade in species and plants, anthropogenic climate change—all these are drivers of biodiversity loss and also drivers of new diseases. Two-thirds of emerging infections and diseases now come from wildlife.[175]

In the last ten thousand years, our hunger for meat and the domestication and exploitation of animals has unleashed untold suffering for all animals—human and nonhuman. Clearly the Meatrix is a driving force in creating the favorable conditions for humans to encounter other novel zoonotic viruses. Unless we change our eating habits, end our love affair with meat, and discontinue trafficking wild and exotic animals, we're likely to see more pandemics in our future.

And zoonosis is not limited to our encounters with wildlife. In fact, Western farming practices also contribute to infectious disease outbreaks. According to a *PNAS* article from May 2013, growing populations, the encroachment by both humans and livestock into wildlife areas, and intensification of agricultural practices together create "opportunities for spillover of previously unknown pathogens into livestock or humans and establishment of new transmission cycles."[176]

For profitability, livestock farming, especially swine and poultry, requires systems in which there are high density and low diversity. The combination of these two factors dramatically increases the risk of pathogenic spillover. In such high-density and low-diversity environments, the authors write, "antimicrobials are often used for growth promotion, disease prevention, or therapeutically, which in turn promotes the evolution of antimicrobial resistance in zoonotic pathogens."[177] A 2020 article published by *The Hill* attributes the following quote to the WHO: "The greatest risk for zoonotic disease transmission occurs at the human-animal interface through direct or indirect human exposure to animals, their products (e.g., meat, milk, eggs...) and/or their environments."[178]

Speaking of high-density and low-diversity factory farming, in October 2021, *ProPublica* published a scathing *exposé* of the US poultry industry's handling—or lack thereof—of an outbreak of a multi-drug-resistant strain of salmonella known as *S. infantis*. The article describes how the government and the chicken industry have knowingly continued to sell contaminated meat that has been making some people sick.[179]

Here's the story: in May of 2018, an outbreak of a virulent strain of salmonella traced to chicken sausages, breasts, and wings made people sick on the East Coast and in geographically diverse areas such as Illinois, West Virginia, the Dominican Republic, and Nicaragua.

Victims wound up in hospitals with stomach pains, uncontrollable diarrhea, and bouts of violent vomiting. Compounding the seriousness of these illnesses was that *Salmonella infantis* is invincible to four of the five medications most commonly prescribed to treat severe food poisoning.

The government and the media have been more than willing to intervene for public food safety and hold industries accountable when non-Meatrix foods are to blame. When a foodborne illness is traced back to plant-based food products, as was the case in the Mexican restaurant Chi Chi's 2003 hepatitis A outbreak linked to green onions from Mexico, and the December 2019 *E. coli* outbreak linked to Fresh Express chopped salad kits, the government stepped in, recalled the products, notified the public, and forced reforms. For instance, due to the FDA's investigation of Fresh Express-brand salad kits, the "Leafy Greens STEC Action Plan" was put into effect to prevent such outbreaks from happening in the future.[180]

Unlike the salmonella chicken outbreak of 2018 that no one heard about during the four years before *ProPublica*'s article, the 2003 Chi-Chi's hepatitis A outbreak was widely reported by the US media, resulting in Chi-Chi's closing all their sixty-five US locations, in part because they were unable to recover from the bad publicity they received. While Chi-Chi's no longer exists in the US, without the stigma of bad press, they still successfully operate restaurants worldwide.[181]

Given the severe nature of the Meatrix's *Salmonella infantis* outbreak, you might expect a similar swift response from the FDA, including warning the public, recalling the tainted poultry, investigating the root cause of the outbreak, and forcing changes within the chicken industry. But the investigation by *ProPublica* indicates none of that happened.

ProPublica's eight-month investigation revealed that while salmonella in poultry plants has been drastically reduced in countries as diverse as Mexico and Sweden, the US lags in consumer safety regarding salmonella in food. The former Head of Food Safety for the World Health Organization, Jørgen Schlundt, who played a vital role in reducing salmonella in Denmark, made this point very clear. During a 2017 roundtable discussion of whole-genome sequencing (touted as the "biggest thing" to happen in food safety in one hundred years), Mr. Schlundt became "increasingly frustrated," observing the cozy relationship between the US government and the US food industry. "I understand that I'm in the US, but surely this must also be about protecting consumers," he told the audience. "We are basically only talking about protecting industry here. I thought that...the basic purpose was to protect consumers, avoid American consumers and other consumers from dying from eating food."[182] Apparently, it's not. So the poultry industry's priority, just like the swine industry before, is in protecting profits, not people.

Clearly, the technology was available to significantly increase food safety prior to the 2018 *Salmonella infantis* outbreak. But because of the close ties between the government and the Meatrix, this same outbreak continues to sicken and kill people. The kind of meat-industry coverup *ProPublica* exposes is simply one more way the Meatrix puts its own bottom line ahead of public safety.

In the previously mentioned *PNAS* article, the authors wrote, "Intensive livestock farming can promote disease transmission through environmental pathways. Ventilation systems expel material, including pathogens such as *Campylobacter* and avian influenza virus, into the environment, increasing risk of transmission to wild and domestic animals." The article also reported that vast quantities of untreated waste from animal agriculture, much of which is spread on land, contaminates watersheds, and comes into contact with wildlife, creating even more pathways for disease transmission.[183]

The Meatrix, and our addiction to animal products, creates these artificial ecosystems where the next pandemic-causing pathogen may emerge. Whether it's wet markets in Asia, Africa, or the Middle East or factory farming in the US and other countries, humans are playing Russian roulette with the next emerging infectious disease—which could easily wind up being more contagious and deadlier than the current coronavirus. The COVID-19 pandemic is our wake-up call, and our response will be vital in determining our future.

What can we do to help prevent infectious diseases from emerging? The short answer is to escape the Meatrix and become plant-based because growing 40 percent of the world's crops to feed livestock (70 percent in the US) fuels biodiversity loss and encroaches into wildlife habitats.[184] We can significantly reduce this by eating plants and living outside the Meatrix.

Hopefully, the world will emerge from the coronavirus pandemic with eyes wide open about the threat that infectious diseases pose to our survival, as well as the ways that we can increase our chances of survival when the next pandemic emerges—and it will.

The saving grace to come out of the coronavirus pandemic might be a greater awareness of the negative consequences of our collective and personal actions. I believe we *can* and *must* make wiser choices when rebuilding our world, leading us to a better, more sustainable, equitable, and compassionate world. Humanity will be better off protecting rather than invading wild animal habitats. We will be better off eliminating wet markets that are breeding grounds for zoonotic diseases. We will be better off growing food for humans, not for animals. We will be better off leaving animals alone to enjoy their lives. We will be better off escaping the Meatrix.

FASCINATING FACTS

- COVID-19 accounted for 74 percent of the increased deaths in the US in 2020.
- A possible origin of the coronavirus pandemic is the pangolin, which is the most trafficked non-human animal in the world.
- 71.8 percent of emerging infectious diseases have a wildlife origin.
- Approximately 500 million people have died of smallpox.
- New-world smallpox killed 80–95 percent of the Indigenous population in North America.
- 200 million people have died from measles. More than 140,000 people died from measles in 2018 alone, most of them children under the age of five.
- 32 million people have died of AIDS.

CHAPTER 4

Escape the Meatrix for the Planet

The scale of the environmental problems we face is over-whelming and makes it very difficult to imagine how one person could make a difference. But one of the best ways to help the environment, lower our carbon footprint, and slow climate change is to escape the Meatrix and adopt plant-based eating. Unfortunately, mainstream news outlets, especially in the US, rarely report this fact. Instead, they tend to focus on the trans-portation sector or green energy solutions like wind and solar. I believe the Meatrix doesn't want our media reporting on their contribution to climate change and that news media often give their Meatrix clients a pass rather than risk losing their lucrative accounts.

One exception to this rule seems to be Britain's newspaper *The Guardian*. In researching this book, I've found their website to contain many articles reporting on the problems of the Meat-rix. For instance, in August 2021, *The Guardian* published an article by Philip Oltermann that discussed the overall trend of Berlin's university students who are leaning toward plant-based eating to combat climate change. In 2010, a vegetarian canteen opened at Berlin's Free University, and a vegan one opened in 2019. Then in 2021, thirty-four canteens and cafes catering to Berlin's four universities began offering only one meat option four days a week. The canteens' remaining fare is 68 percent vegan, 28 percent vegetarian, and 2 percent fish-based. For the city's Humboldt University and Berlin's Technical University, this effort is part of a commitment to become carbon neutral by 2030 and 2045, respectively.[185]

These are encouraging signs that connections are being made between the food on our plate and climate change despite the lack of widespread reporting. But the Meatrix prefers you don't make these connections at all. In fact, the Meatrix knows it has a climate-change problem, and rather than tackling its emissions problems head-on, it works hard to prevent the passage of environmental laws that would restrict its activities.

The Meatrix spent almost $200 million from 2000 to 2019 lobbying against environmental and climate-change regulations in the US. That's roughly 10.5 million per year![186] I believe this money would have been better spent helping farmers retool their operations for a plant-based food economy.

Cutting emissions from agriculture is a critical component of mitigating runaway climate change, but more than half of the largest livestock-producing countries fail to even mention reducing animal agriculture as part of their goals for reaching the targets of the Paris Climate Accords.[187]

Ignoring the many ways the Meatrix contributes to climate change is a big problem, but things may be looking up.

For instance, a 2019 study by Oxford University found that the best dietary pattern for individuals wishing to reduce their carbon footprint was plant-based. By analyzing 38,700 farms in 119 countries, the study concluded that people in the US, where meat consumption per capita is three times the global average, could lower their carbon footprint by a whopping 73 percent alone by switching to a plant-based lifestyle. This impact is "far bigger than cutting down on your flights or buying an electric car."[188]

Additionally, consuming only plant-based foods does more for the environment than simply lowering greenhouse gases; it also benefits the earth by reducing global acidification, land and water use, habitat and biodiversity loss, and the mass extinction of wildlife. The impact was so impressive and conclusive that the lead Oxford scientist gave up eating meat and dairy altogether one year into the five-year study.[189]

Despite the Meatrix's intense lobbying efforts and a lack of widespread reporting, information about how the Meatrix contributes to climate change *is* available if you choose to seek it out. Many people have read reports of a study done by the United Nations' Food and Agricultural Organization (FAO) saying that animal agriculture accounted for 14.5 percent of human-generated greenhouse gases.[190] Few people realize the original 2006 study, "Livestock's Long Shadow," estimated that number to be 18 percent.[191]

Even though the media frequently cites the 14.5 percent figure, not all scientists agree the number is that low. For example, in April 2019, Steven Chu, president of the American Association for the Advancement of Science, former Energy Secretary, and Nobel Prize recipient, reported the amount of greenhouse gases generated by the Meatrix was much higher. "If cattle and dairy cows were a country, they would have more greenhouse gas emissions than the entire E.U. 28," said Chu. He also reported that the previously mentioned FAO report omits two major contributors while underestimating another. When we account for everything, Chu says the total greenhouse gases generated by animal agriculture is 51 percent, exceeding the greenhouse gases from power generation.[192]

For instance, the FAO's report entirely excludes greenhouse gases generated by both animal respiration and farmed fish. This is problematic because ruminant animals have a multi-chambered stomach, called a rumen, that allows them to digest the cellulose found in the grass they consume through a process called enteric fermentation. Unfortunately, one of the byproducts of enteric fermentation is methane (CH_4).

While grass-fed beef is considered by some to have low environmental impact, it currently accounts for only 4 percent of beef in the US. Still plant-based foods have a much lower impact than grass-fed beef. "Converting grass into [meat] is like converting coal to energy. It comes with an immense cost in emissions," according to Dr. Joseph Poore, lead author of the previously mentioned five-year Oxford University study.[193]

According to the Environmental Protection Agency (EPA), methane is a greenhouse gas that remains in the atmosphere for about twelve years, after which 80 percent to 90 percent of it is naturally removed. However, the EPA website also says, "Methane's lifetime in the atmosphere is much shorter than carbon dioxide (CO_2), but CH_4 is more efficient at trapping radiation than CO_2. Pound for pound, the comparative impact of CH_4 is 25 times greater than CO_2 over a 100-year period." Additionally, the same EPA article states, "When livestock and manure emissions are combined, the Agriculture sector is the largest source of CH_4 emissions in the United States."[194]

The following graphs compare the short- versus long-lived greenhouse gas emissions from various foods. The first one is of a variety of foods, while the second one focuses on protein-rich foods only. These graphs are potent visual reminders of the Meatrix's impact on climate change and why shifting to plant-based foods is such a powerful tool in combating climate change.

The exclusion of farmed fish in the FAO report also seems shortsighted to me because a 2018 article by Sustainable

GREENHOUSE GAS EMISSIONS FROM FOOD, SHORT- VS. LONG-LIVED GASES

Greenhouse gas emissions are measured in carbon dioxide equivalents (CO$_2$eq) based on their 100-year global warming potential (GWP). Global mean emissions for each food are shown with and without the inclusion of methane—a short-lived but potent greenhouse gas

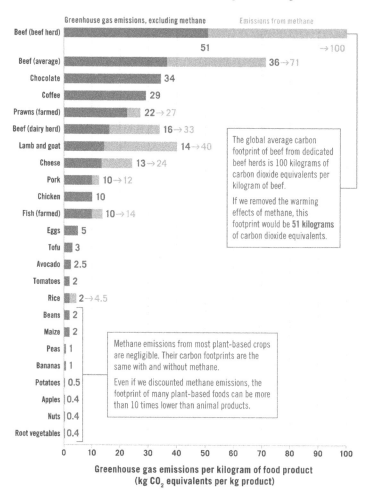

Greenhouse gas emissions, excluding methane   Emissions from methane

Food	Value
Beef (beef herd)	51 →100
Beef (average)	36→71
Chocolate	34
Coffee	29
Prawns (farmed)	22→27
Beef (dairy herd)	16→33
Lamb and goat	14→40
Cheese	13→24
Pork	10→12
Chicken	10
Fish (farmed)	10→14
Eggs	5
Tofu	3
Avocado	2.5
Tomatoes	2
Rice	2→4.5
Beans	2
Maize	2
Peas	1
Bananas	1
Potatoes	0.5
Apples	0.4
Nuts	0.4
Root vegetables	0.4

The global average carbon footprint of beef from dedicated beef herds is 100 kilograms of carbon dioxide equivalents per kilogram of beef.

If we removed the warming effects of methane, this footprint would be **51 kilograms** of carbon dioxide equivalents.

Methane emissions from most plant-based crops are negligible. Their carbon footprints are the same with and without methane.

Even if we discounted methane emissions, the footprint of many plant-based foods can be more than 10 times lower than animal products.

Greenhouse gas emissions per kilogram of food product
(kg CO$_2$ equivalents per kg product)

Source: Hannah Ritchie and Max Roser, "Environmental Impacts of Food Production," *Our World in Data* (2020), ourworldindata.org/environmental-impacts-of-food

GREENHOUSE GAS EMISSIONS FROM PROTEIN-RICH FOODS, SHORT- VS. LONG-LIVED GREENHOUSE GASES

Greenhouse gas emissions are measured in carbon dioxide equivalents (CO_2eq) based on their 100-year global warming potential (GWP). Global mean emissions for each food are shown with and without the inclusion of methane—a short-lived but potent greenhouse gas

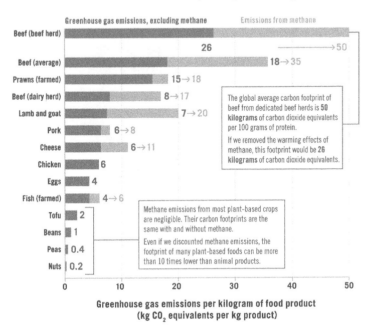

Greenhouse gas emissions per kilogram of food product
(kg CO_2 equivalents per kg product)

Source: Hannah Ritchie and Max Roser, "Environmental Impacts of Food Production," *Our World in Data* (2020), ourworldindata.org/environmental-impacts-of-food

Fisheries, funded by the University of Washington, found that farmed catfish, shrimp, and tilapia can use greater amounts of energy than livestock, depending on the source of the electricity used for water circulation and whether it comes from fossil fuels or climate-friendly sources such as solar.[195] In addition, a 2019 *European CEO* article found that "although in many cases farmed seafood has a lower carbon footprint than, say, beef, this is not always the case: for the same weight of protein, catfish farming actually produces more greenhouse gases than cattle farming, while shrimp, tilapia and carp are not far behind."[196]

But it's not only the source of the electricity used in water circulation that can be a major contributor to greenhouse gases in farmed fish. "You get all these fish depositing excreta and unconsumed feed down to the bottom of the pond, where there is barely any oxygen, making it the perfect environment for methane production," a potent greenhouse gas, says Dr. Joseph Poore.[197] That's a lot of methane, considering that 66 percent of fish consumed in Asia and 96 percent in Europe are farmed fish.

The FAO report also included scientifically unproven items such as carbon sequestration by livestock (the belief that grazing cattle can help offset global warming by stimulating soil to absorb more carbon from the atmosphere). But even during optimal conditions (perfect soil moisture content and the optimal number of animals per field), carbon sequestration by livestock is not adequate to counteract ruminant emissions.[198] In fact, using improved pasture management accounts for a maximum 22 percent reduction in ruminate emissions, according to a comprehensive 2018 Oxford University study.[199]

Even if the opposite *were* true, carbon sequestration in soil is not a permanent solution. An article by the Institute for Agricultural & Trade Policy, a long-standing, independent agricultural/trade organization, stated the problem with carbon storage is that it's "extremely impermanent" because many no-till farmers till at least once every few years as a way to manage weeds. Doing so releases much of the stored carbon back into the atmosphere. The article continued, "Even long-term contracts that bind land managers to use certain practices do not ensure permanence since the carbon stored can be released back into the atmosphere as soon as the contract is up if the land manager returns to less climate-friendly practices."[200]

These three issues—enteric fermentation, farmed fish, and carbon sequestration—bring the FAO's assessment of animal agriculture's carbon footprint into question.

Getting back to Steven Chu's assessment that animal agriculture contributes more to global warming than power generation, it turns out he's not alone. In their 2009 paper, Robert Goodland, a former lead environmental advisor at the World Bank Group, and Jeff Anhang, a research officer and environmental specialist at the World Bank Group's International Finance Corporation, reported that "livestock and their byproducts actually account for at least 32,564 million tons of CO2e (carbon dioxide equivalent) per year or 51 percent of annual worldwide GHG emissions."[201]

Whether 14.5 percent, 51 percent, or somewhere in between, it's risky for us to ignore animal agriculture's large carbon footprint. When combined with other significant concerns, such as land and water degradation and biodiversity loss, it's clear the Meatrix is creating an environmental catastrophe.

In fact, in 2016, University of Oxford scientists using computer modeling determined that food-related emissions would drop by 60 percent if everyone became vegetarian by 2050. Most of these reduced emissions would result from simply eliminating red meat. Computer modeling further indicated the reduction would rise to 70 percent if everyone became plant-based.[202]

The same report also validates many of the previous section's health claims by concluding that adopting a plant-based lifestyle could prevent 8.1 million deaths *per year* by 2050. Lead author of the study Dr. Marco Springmann said, "Approximately half of the avoided deaths were due to reduction of red meat consumption, with the other half due to a combination of increased fruit and vegetable intake and a reduction in calories, leading to fewer people being overweight or obese."[203]

As I write this in December of 2021, the John Hopkins Coronavirus Resource Center reports that COVID-19 is responsible for more than 818,000 US deaths and 5.4 million deaths worldwide.[204] With enormous effort and financial investment, governments created vaccines in record time. Vaccination campaigns to protect people's health and avoid unnecessary deaths are proving successful. Nevertheless, current worldwide COVID-19 deaths are about 66 percent of the 8.1 million deaths per year that the Oxford Martin Programme on the Future of Food projected would be saved annually if we all adopted a plant-based lifestyle. The coronavirus pandemic has been in the news daily, with people and businesses making massive changes to avoid further deaths and economic disruptions. It makes me wonder how many more lives could be saved if our media gave studies like the University of Oxford's as much airtime as the pandemic.

Not only might it save your life, but eating plants is a much more sensible use of resources than growing plants that feed livestock to feed humans. For example, it takes "2,500 gallons of water, 12 pounds of grain, 35 pounds of topsoil and the energy equivalent of one gallon of gasoline to produce one pound of feedlot beef."[205]

According to a frequently cited report by David Pimentel, a professor of ecology in Cornell University's College of Agriculture and Life Sciences, the grain the US alone feeds to livestock could feed almost eight hundred million people. Pimentel's 1997 report estimates that forty-one million tons of plant protein are fed to US livestock each year to generate only seven million tons of consumable animal protein. When comparing the fossil fuel input to the animal protein output, beef has the highest ratio at fifty-four to one.[206]

A 2013 study that attempted to redefine crop yields from tons of food produced to the number of people nurtured per hectare revealed even more astonishing statistics about animal agriculture's inefficiencies. According to the study, growing crops exclusively for human, rather than animal consumption, would generate a 70 percent increase in available calories, or feed an additional four billion people. The study's authors write, "The US agricultural system alone could feed 1 billion additional people by shifting crop calories to direct human consumption." The study continues to claim that even small shifts in crops grown for animal agriculture and biofuels could positively impact global food security.[207]

Decidedly, animal agriculture is not a good use of land, water, and resources. A powerful way to help mitigate climate change, improve our environment, and conserve our natural resources would be to raise crops for humans instead of livestock.

According to another article, chickens have the best feed conversion ratios (FCRs) in animal agriculture. Here's a list of the FCR of edible weight for the leading animal protein sources:

- Beef cattle: 25:1
- Pigs: 9.4:1
- Chickens: 4.5:1[208]

To illustrate these inefficiencies: Would you invest money in the stock market (or anywhere else) if you knew that for every $450 you invested, the best you could ever hope to get back was $100 (losing $350 in the process)? Of course not. You would think that was a pretty horrible return on investment (ROI). But this is precisely the pitiful ROI we're getting from chickens.

In addition, it takes almost one hundred times more land to produce a single gram of protein from beef or lamb than from tofu.[209]

As populations continue to increase, our ability to feed everyone is a significant concern. According to one article, "The research suggests that it's possible to feed everyone in the world a nutritious diet on existing croplands, but only if we saw a widespread shift towards plant-based diets."[210]

LAND USE OF FOODS PER 1,000 KILOCALORIES

Land use is measured in meters squared (m²) required to produce 1000 kilocalories of a given food product.

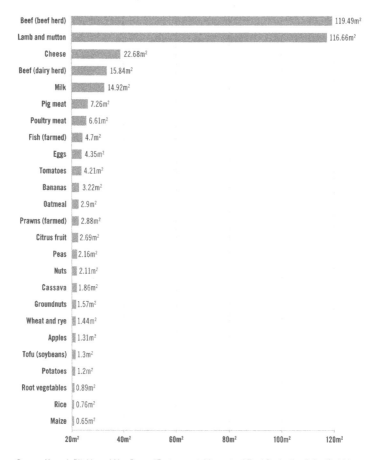

Food	Land use
Beef (beef herd)	119.49m²
Lamb and mutton	116.66m²
Cheese	22.68m²
Beef (dairy herd)	15.84m²
Milk	14.92m²
Pig meat	7.26m²
Poultry meat	6.61m²
Fish (farmed)	4.7m²
Eggs	4.35m²
Tomatoes	4.21m²
Bananas	3.22m²
Oatmeal	2.9m²
Prawns (farmed)	2.88m²
Citrus fruit	2.69m²
Peas	2.16m²
Nuts	2.11m²
Cassava	1.86m²
Groundnuts	1.57m²
Wheat and rye	1.44m²
Apples	1.31m²
Tofu (soybeans)	1.3m²
Potatoes	1.2m²
Root vegetables	0.89m²
Rice	0.76m²
Maize	0.65m²

Source: Hannah Ritchie and Max Roser, "Environmental Impacts of Food Production," *Our World in Data* (2020), ourworldindata.org/environmental-impacts-of-food

Not only is agricultural land use an important consideration for environmentalists, the overall composition, structure, and integrity of the earth's biosphere is of equal concern.

For the first time, in 2018, scientists estimated a comprehensive, holistic approach to determining the biomass of the planet.

These scientists established a consensus of the earth's biomass at approximately 550 cubic gigatons of carbon (Gt C) distributed among all kingdoms of life. The study determined the combined biomass of the 7.6 billion humans on the planet to be 0.06 Gt C. Humans are but one species comprising a minuscule amount of the earth's total biomass.[211] Still, we have altered the entire world by our activities.

For instance, according to one study, the biomass of wild land mammals has decreased 85 percent since the rise of human civilization. Today, farmed poultry comprises 70 percent of all birds on the planet, with only 30 percent being wild.[212]

According to many scientists, we are now witnessing the sixth mass extinction crisis on the planet. Extinction is a natural phenomenon that typically happens at a background rate of one to five species per year. However, the late E. O. Wilson, former Harvard biologist and one of the world's foremost proponents of biodiversity, in 1993 estimated today's extinction rates are occurring at thirty thousand species per year, or three per hour.[213] So you can see why scientists consider this a crisis of epic proportions.

In 2019, the United Nations Intergovernmental Science-Policy Platform on Biodiversity and Ecosystem Services (IPBES) issued a press release estimating that extinction now threatens one million species. The study is considered the most comprehensive to date, involving 145 expert authors and input by another 310 authors from 50 countries.[214]

Unlike other cataclysmic extinction events caused naturally by volcanoes or asteroids hitting the earth, human activity is responsible for today's extinction rates.

Remember, a vast majority of the accelerated extinctions stem from humans, whose biomass comprises only one-hundredth of a percent of the Earth's 550 Gt C biomass.

Sadly, humans have always had a profound impact on the

planet. Extinctions followed when *Homo sapiens* migrated from Africa into Asia, Europe, Australia, North America, and the Caribbean.[215] So not only is human activity the cause of the crisis we now face, but we're also likely to be its victim.

To look at it another way, extinction rates since 1980 are 165 times greater than the rates seen during the most recent mass extinction event known as the Cretaceous-Paleogene (K-Pg), which killed off the dinosaurs. Additionally, it's highly likely that many species we currently consider "threatened" will go extinct within the next one hundred years. When we include these "threatened" species, today's extinction rates are thousands of times faster than those seen during the K-Pg.[216]

HOW MANY TIMES FASTER ARE SPECIES GOING EXTINCT RELATIVE TO THE CRETACEOUS MASS EXTINCTION EVENT?

Recent extinction rates are compared to estimated rates from the fossil records of the Cretaceous-Palogene (K-Pg) mass extinction event, 65 million years ago.

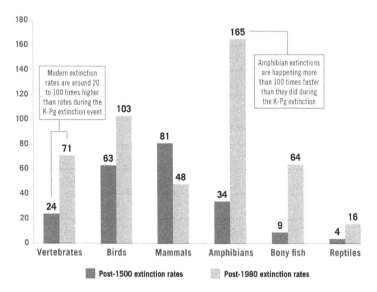

Source: Hannah Ritchie and Max Roser, "Biodiversity," *Our World in Data* (2021), ourworldindata.org/extinctions

In January 2019 the medical journal *The Lancet* published an article by its Lancet Commission, composed of eighteen commissioners and eighteen coauthors from sixteen countries in various fields, including human health, agriculture, environmental sustainability, and political sciences. The commission's goal was to make sustainable recommendations for food production and human health.[217]

Based on the objectives of the UN Sustainable Development Goals, and the Paris Climate Accords, the commission defined global targets of diet and food production. The commission's recommended diet consists of vegetables, fruits, whole grains, legumes, nuts, and unsaturated oils, with zero or low quantity red meat, processed meat, added sugar, refined grains, and starchy vegetables.

Number one on their list of key messages is 820 million people are food insecure and malnourished, while others consume an unhealthy diet contributing to premature death and morbidity. The commission went further, saying that "global food production is the largest pressure caused by humans on Earth, threatening local ecosystems and the stability of the Earth system."[218]

I think this information would be shocking to many people. If asked, most people would probably say energy production (the fossil fuel industry or coal mining) or the transportation sector (planes, trains, and automobiles) puts more pressure on the Earth than food production. They'd be wrong, according to this study.

Except for people in sub-Saharan Africa, who are among the most nutritionally insecure on the planet, the commission's recommendations for red meat, poultry, fish, and dairy begin at zero grams per day. Eggs are the only Meatrix product that was not recommended to start at zero grams per day. The report says replacing a starchy vegetable with one egg per day may reduce stunting in malnourished children, a condition found in some developing countries, particularly in South Asia (which in

2018 accounted for 38.9 percent of the world's growth-stunted children).[219] Moreover, while the commission recommended an intake of one to five eggs per week, it went on to say isocaloric (having the same or similar caloric values) plant-based substitutions for eggs might reduce the risk of noncommunicable diseases. The takeaway is that unless you're suffering from malnutrition, eggs offer no health protection and, in fact, consuming them may increase your risk of disease, particularly heart disease in people with diabetes.[220]

At this point, you may be wondering exactly how the Meatrix contributes to mass extinction rates and threatens local ecosystems. Today, as mentioned in Chapter 1, most farm animals are raised on factory farms and fed a diet primarily consisting of wheat, corn, and soy.[221] In the United States alone, over 260 million acres of forest have been clear-cut to grow crops for livestock.[222] We fuel extinction rates when we clear-cut a diverse environment and replace it with a monoculture like wheat, corn, or soy.

The extinction rates are even higher in tropical areas with more biodiversity, such as the Amazon rainforest. Sadly, in 2020, the Brazilian Amazon deforestation rate was the greatest of the past decade, 182 percent higher than the target rate set by Brazil's National Policy on Climate Change.[223]

Clearly, countries such as Brazil have been unable to meet their commitments to biodiversity and deforestation reductions, but how can individuals take a stand for biodiversity and climate change? There are many ways to lower your carbon footprint, like using public transportation, riding a bike to work, buying local, and taking shorter hot showers. These are all excellent choices. But one of the most effective ways is to become plant-based.

If we eat only organic foods, we're lowering our footprint even more by reducing the use of synthetic fertilizers that drive up global emissions of nitrous oxide (N_2O), a lesser-known greenhouse gas. "As a climate pollutant, N_2O can linger in the atmosphere for decades, and is far more efficient than CO_2 in trapping heat."[224]

I ran across a poster created by simplehappykitchen.com containing the following information:

EVERY DAY A VEGAN SAVES

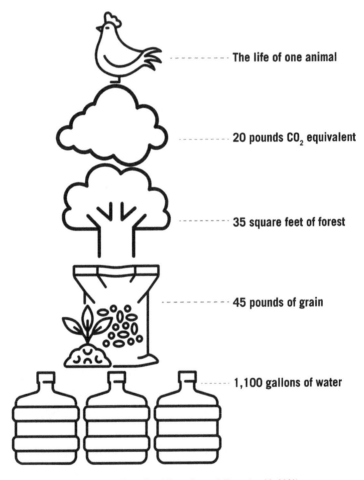

The life of one animal

20 pounds CO_2 equivalent

35 square feet of forest

45 pounds of grain

1,100 gallons of water

Source: Simple Happy Kitchen, "Every Day A Vegan Saves..." (December 20, 2021), simplehappykitchen.com/vegan-news/every-day-a-vegan-saves

I was simultaneously encouraged and suspicious. Could my food choices be having such a dramatic impact daily? Upon further inquiry, I discovered the "Every Day a Vegan Saves" infographic uses government publications for its figures. Talk about making a *huge* difference for the environment! Whenever I feel overwhelmed about climate change, biodiversity loss, and the current extinction rates, I think of this poster and remind myself that even though it might seem as though I'm not doing much, thanks to my plant-based eating, I'm actually doing quite a bit.

The first chart that follows shows the benefits of eating only plant-based foods. It lists food items into four food groups (plant-based, fish, dairy/eggs, and meat) and puts their environmental effect into perspective. When you compare each food to *The Lancet* Commission's five environmental criteria, you can see that Meatrix products lead the charge in destroying our planet.[225]

The environmental costs of eating within the Meatrix are higher, by far, than the typical consumer could know. Although currently, a McDonald's Quarter Pounder with Cheese seems like a bargain at the cost of only $3.79, the actual cost to the environment is considerable. For instance, the Water Footprint Network estimates it takes more than 465 gallons of water on average to produce one-quarter pound of beef.[226]

The second chart that follows compares the water requirement per ton of food product and shows that six of the top eight most water-intensive agricultural products are within the Meatrix.[227]

For example, beef requires almost twice the water as its nearest competitor, sheep/goat meat. Number two on the list is nuts, which is why some environmentally conscious consumers avoid nut milks.[228] As far as plant-based nut milks go, almond milk is the most water-intensive of all the plant-based milks but generates fewer greenhouse gases because the almond trees sequester so much carbon over their lifetime. The least water-intensive plant-based milk is soy. A 200-milliliter glass of

ENVIRONMENTAL EFFECTS OF DIFFERENT FOOD GROUPS, PER SERVING

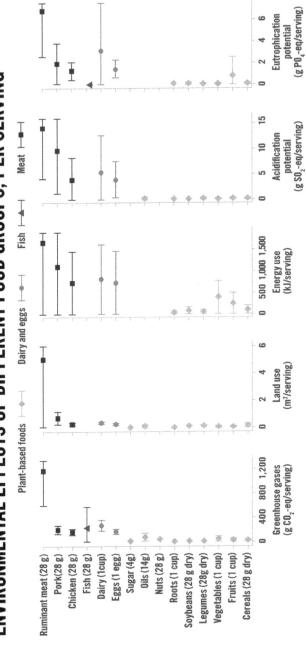

Source: Michael A. Clark, et. al. "Multiple health and environmental impacts of foods," *Proceedings of the National Academy of Sciences* 116, no. 46 (October 2019), doi.org/10.1073/pnas.1906908116

WATER REQUIREMENT PER TON OF FOOD PRODUCT

Global average water footprint of food production, which includes water requirements across its full supply chain and the quantity of freshwater pollution as a result of production.

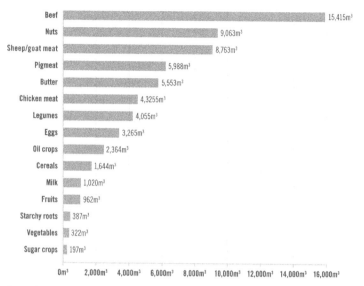

Beef	15,415m³
Nuts	9,063m³
Sheep/goat meat	8,763m³
Pigmeat	5,988m³
Butter	5,553m³
Chicken meat	4,3255m³
Legumes	4,055m³
Eggs	3,265m³
Oil crops	2,364m³
Cereals	1,644m³
Milk	1,020m³
Fruits	962m³
Starchy roots	387m³
Vegetables	322m³
Sugar crops	197m³

0m³   2,000m³   4,000m³   6,000m³   8,000m³   10,000m³   12,000m³   14,000m³   16,000m³

Source: Hannah Ritchie and Max Roser, "Environmental Impacts of Food Production," *Our World in Data* (2020), ourworldindata.org/environmental-impacts-of-food#water-use

soymilk requires 54 liters of water. By comparison, dairy milk uses 125.6 liters.[229] The water footprint of the Meatrix is immense.

It takes thirteen pounds of feed and sixty-five square feet of land to create a quarter pound of beef, while generating about four pounds of greenhouse gases. *Business Insider* put it this way: "In reality, your hamburger isn't something bought and paid for, but a symbol of a debt that, one day, must be repaid."[230]

The idea of a burger symbolizing a debt reminded me of a significant environmental quote by Kentucky Hall of Fame writer Wendell Berry. In his book, *The Unforeseen Wilderness: An Essay on Kentucky's Red River Gorge*, Mr. Berry wrote that "the world is not given by his fathers, but borrowed from his children." The choices each of us makes now will determine the amount of debt owed. Everyone living within the Meatrix fuels greater and greater climate change, all but guaranteeing that their children and children's children will inherit a debt so vast that no one could possibly ever repay it.

After learning about the horrific environmental impact of beef, I wondered how my favorite go-to, plant-based burger—Beyond Meat's Beyond Burger—stacked up. I was delighted to find a report by the Center for Sustainable Systems at the University of Michigan revealing that compared to a quarter-pound of US beef, the Beyond Burger generates 90 percent fewer greenhouse emissions, requires 46 percent less energy, has 99 percent less impact on water scarcity, and has 93 percent less impact on land use.[231] The reduced environmental impact of the Beyond Burger is something that everyone can celebrate. And, as many surprised meat eaters will tell you, the Beyond Burger tastes fantastic!

We're shattering climate-related records at an alarming rate. For example, on September 2, 2021 the Central Park Automated Surface Observing System recorded 3.15 inches of rainfall in one hour from Hurricane Ida, breaking a record set by Hurricane Henri only eleven days prior.[232] On the same date, New York City's "Midtown saw 7.49 inches while Staten Island recorded 8.92 inches, according to the National Weather Service."[233]

This weakened storm left dozens dead, including at least eleven people who drowned in their basement apartments, unable to get out as quickly as the water was rushing in. In addition, the flooding shut down the transit system and resulted in canceled flights. Sadly, such storms are becoming more frequent as the earth heats up.

Speaking of heating up, in September 2021, Weather.com reported temperatures for summer 2021 were 2.6 degrees above average and that four of the five hottest summers for the contiguous US had occurred in the past eleven years.[234]

We're seeing record-breaking rainfall and increased flooding in some areas of the planet while others suffer never-before-seen heatwaves, droughts, and forest fires. Climate change is here. Period. We're all borrowing this world from future generations, and we owe it to them to do our part to mitigate the damage. Escaping the Meatrix and going plant-based is the number one action each of us can take.

As we've seen, our individual and collective dietary choices have a significant impact on our health, the environment, and climate change. Unfortunately, we forget how interconnected and reliant we are on nature when we are working surrounded by concrete, living in buildings in highly populated areas, buying our food in grocery stores, and eating at restaurants. As a result, we often consider ourselves as somehow separate from the natural world. Yet we can learn a great deal from observing the natural world around us. Nature provides us with powerful clues about what and how we should be eating, and I would like to mention two in this section.

The first concerns our consumption of dairy products, which the Meatrix leads us to believe is the most natural thing in the world. However, our dairy consumption is contrary to everything we see in nature. There are more than 5,400 mammals on earth[235], yet we are the only species that drinks the milk created for the baby of a different species.[236]

A cow's milk is designed specifically for her calf, with all the nutrients it needs to grow strong and healthy. In my mind, it's kind of perverted for us to steal the milk designed to speed the development of a baby calf until it can live on solid food. Forgoing dairy would not only end the enormous suffering mother cows and their babies endure (see Chapter 5), but it would also significantly reduce animal agriculture's environmental footprint and help us improve our health.

Calcium is good for building strong bones, and the Meatrix would like us to believe it has a monopoly on this nutrient. But the cow doesn't produce calcium at all. Calcium is found in cow's milk because it's in the earth and absorbed by the roots of the plants that cows eat. However, only 4 percent of the US beef market is grass-fed. The other 96 percent of cattle requires supplementation with calcium. Knowing cows are not the source of calcium means we can cut out the middle cow (so to speak) and get our calcium in the same way 96 percent of cows today get theirs: from the earth.[237] Many plant foods provide sources of calcium: chia seeds, lentils, beans, almonds, collard greens, spinach, kale—the list goes on and on. Our milk consumption should end when we are weaned from our mother's milk, the natural process we see in every other species. Leave the cows' milk for baby cows.

The second clue has to do with the type of foods nature intends for us to eat. Many people believe our natural diet is omnivorous. However, eating animal-based foods is something we're taught to do and conditioned to think of as normal. It turns out, eating animal products is not a requirement of human physiology. In fact, a good litmus test for whether we are designed to eat meat is how we respond when we see roadkill. Do we avert our eyes, or do we pull over and have a snack? If we stopped and scavenged, we would become deathly ill. Watching slaughterhouse footage usually disturbs or even nauseates us. If we were carnivorous, we would salivate at the sight of blood. Watching animals die would make us hungry, and consuming roadkill would not make us ill.

Another indicator that we are herbivores by nature is our position on the global food chain. We like to think that we're at the top, but that is wrong. The hierarchical position an organism occupies in a food web is known as its trophic level. Each level represents how many consumption steps exist between an organism and the food chain's original energy source, the sun.

Level 1 is plants that get their nutrients directly from the sun. Level 2 is herbivores that get their nutrients from plants. Level 3 is carnivores that get their nutrients from herbivores. Level 4 is carnivores that eat other carnivores. And level 5 is apex predators with no predators. Level 5 is at the top of the food chain—like polar bears, saltwater crocodiles, and orca whales. We like to think of ourselves as sitting at the top of all food chains, but humans actually have a trophic level of 2.2. The lead researcher at the French Research Institute for Exploitation of the Sea, Sylvain Bonhommeau, said, "We are closer to herbivore than carnivore. It changes the preconception of being top predator."[238]

Also, according to the UN, in 2019 80 percent of what we ate globally came from plants.[239] But as people become more affluent, their preferences shift to a more meat-centric diet.[240]

As mentioned earlier, while the US is seeing a rise in plant-based eating, the world as a whole is becoming more carnivorous. The growing global appetite for meat is fantastic news for the Meatrix. But this expansion of the Meatrix is not only devastating for the environment; it's also deadly for the animals and terrible for our health. I think it's important to keep in mind that the shift from plant-based eating to a more animal-based diet is not biological. It's a matter of preferences for animal products—not only their taste but their association with Western culture—overruling our nature.

Other physiological traits indicate humans were meant to be herbivores. For example, humans' canine teeth are more or less the same length as our other teeth. By comparison, if you examine the canine teeth of a carnivore, such as a mountain lion, bear, or even the Nile crocodile, you'll notice that their canines are much sharper and extend considerably farther than the rest of their teeth. More extended and sharper canines allow carnivores to rip and tear flesh and swallow it in whole chunks. Furthermore, the human jaw can move up and down *and* sideways—whereas a carnivore's jaw moves only up and down. And like other herbivores, a human's molars are flat for grinding plant material.

In addition to not having a carnivore's teeth and jaw structure, humans' bodies are ill-equipped to kill prey: we lack sharp claws to catch and kill animals. This is why humans have invented weapons that allow us to do so.

An essential role for the stomach acids of carnivores is to break down the protein and kill the bacteria within the flesh they consume. Humans have comparatively weaker stomach acid, predisposing us to break down pre-chewed plant material.

Carnivorous animals have a short digestive tract that allows the meat they consume to pass more quickly through their bodies so they don't become sick from rotting flesh in their intestines. Our digestive tracts are much longer than those of comparably sized carnivores. Our longer intestines allow us to break down the fiber and absorb the nutrients from fruits and vegetables. The lengthy time it takes for meat to pass through our digestive system is one of the reasons that eating animal flesh is so bad for us. In fact, "Proteins in meat and fish can take as long as two days to fully digest, while fruits and vegetables may move through your system in less than a day due to the higher fiber content."[241]

When asked why our ancestors sometimes resorted to eating meat, Briana Pobiner, a paleoanthropologist at the Smithsonian's National Museum of Natural History, says:

> Fruit and different plants and other things that we may have eaten maybe became less available...the meat-eating that we do, or that our ancestors did even back to the earliest time we were eating meat, is culturally mediated. You need some kind of processing technology in order to eat meat...so I don't necessarily think we are hardwired to eat meat.[242]

I believe our physiology agrees. Just because we *can* do some-

PHYSIOLOGICAL TRAITS OF CARNIVORES, OMNIVORES, HERBIVORES, AND HUMANS

	Carnivore	Omnivore	Herbivore	Human
Facial muscles:	Reduced	Reduced	Well developed	Well developed
Mouth opening:	Large	Large	Small	Small
Teeth:	Sharp	Sharp	Flat	Flat
Saliva:	No enzyme	No enzyme	Carbohydrate enzyme	Carbohydrate enzyme
Liver:	Cannot detoxify vitamin A	Cannot detoxify vitamin A	Can detoxify vitamin A	Can detoxify vitamin A
Small intestine:	3-6 times body length	4-6 times body length	10-11 times body length	10-11 times body length
Urine:	Acidic	Acidic	Alkaline	Alkaline
Stomach PH:	1	1	4-5	4-5

Source: Dr. Milton Mills, "The Comparative Anatomy of Eating," *Dr. Milton Mills Plant-Based Nation* (November 15, 2019), drmiltonmillsplantbasednation.com/the-comparative-anatomy-of-eating/, toppr.com/guides/biology/difference-between/herbivores-and-carnivores/

thing doesn't mean we *should* do it. For example, we can eat animal products and live, but are we thriving by doing so? When we eat animal foods, are we fueling our bodies as nature intended? I would say no, and I think science agrees. So the argument that we can eat both animals and plants, and therefore it's okay to eat animals, has no scientific basis.

FASCINATING FACTS

- US agribusinesses spent $200 million between 2000–2019 lobbying against climate change.

- Being plant-based results in a 73 percent reduction in one's personal carbon footprint.
- If cattle and dairy cows were a country, their emissions would exceed those by the combined 27 countries in the EU plus the United Kingdom.
- Animal agricultural is the source of 51 percent of all greenhouse gas emissions.
- The comparative impact of methane is 25 times greater than CO_2 over a 100-year period.
- Of the twelve foods that produce the most greenhouse gas emissions, ten are animal-based. The only two that are not are chocolate and coffee.
- Global greenhouse gases would drop by 70 percent if everyone became plant-based.
- 8.1 million deaths would be avoided each year by 2050 if everyone became plant-based.
- The ratio of fossil fuel input to the animal protein output of beef is 54:1.
- 96 percent of all protein in cereal and leguminous grains fed to animals is not converted to edible protein.
- Growing crops exclusively for human consumption could feed an additional 4 billion people.
- It takes almost 100 times more land to produce a gram of protein from beef or lamb than to produce tofu.
- Farmed poultry comprises 70 percent of all birds on the planet
- A plant-based eater saves 1,100 gallons of water, 45 pounds of grain, 35 square feet of forests, 20 pounds of CO_2 equivalent, and the life of one animal—every day
- Beyond Meat's Beyond Burger generates 90 percent fewer greenhouse emissions, requires 46 percent less energy, has 99 percent less impact on water scarcity, and has 93 percent less impact on land use than beef.

CHAPTER 5

Escape the Meatrix for the Animals

We're uncovering the many ways the Meatrix negatively impacts our world. So far, we've discussed how escaping the Meatrix improves our own health, the health of the planet, as well as reduces the potential for new and deadly emerging infectious diseases. We've also learned how the Meatrix contributes to environmental degradation and fuels climate change.

In this chapter we'll explore three more reasons to escape the Meatrix. These are animals, humans, and spirituality.

ANIMALS

I was running along a tree-lined path on the northeast shore of Oahu in May 2019. I was in Hawai'i for business and decided to squeeze in some exercise with a late-afternoon run. As I navigated a blind turn in the path, I saw a young man and woman sitting under a canopy doing what appeared to be homework. On my return back to my hotel, I stopped to greet them.

The young woman told me they were marine biology majors at the University of Hawai'i and participating in a summer internship with the National Oceanic and Atmospheric Administration (NOAA). A Hawaiian monk seal had given birth to a pup in the protected waters of the small cove in front of us. The students told me they were there to observe and document the seals' activities. NOAA cordoned off a section of the beach to prevent tourists and locals from disturbing the mother and her baby. She said the seals are native to Hawai'i and give birth to their babies in May. Over the next few days, I went back to that beach to watch the monk seal, known as Honey Girl, and her baby. On one of those afternoons, I had an interesting conversation with the woman.

I asked how the Hawaiian monk seal had become so endangered. She explained a prolonged decline began in the 1950s, but things were starting to stabilize due to efforts to save them. She continued that some of the threats to monk seals were lack of prey, shark attacks, and habitat loss but that the greatest danger was discarded fishnets, a practice NOAA was working to end.

We chatted about the plight of the monk seal and the overall polluted state of the planet's oceans. During our conversation, we discussed how we could help save the Hawaiian monk seal and all marine life by simply not eating seafood. If we didn't eat fish, there would be no financial incentive for people to fish, meaning fewer discarded nets. I shared that it's possible to live without eating any animal products and thrive while doing so. I mentioned plant-based eating, and the challenge22.com website to her. She said she had heard of it and would check it out. I told her I thought if she went plant-based, she would likely look back at it as one of the best decisions of her life. I thanked her for our conversation, said goodbye, and headed back to my hotel, happy I'd met someone dedicated to making the planet a safer place for Hawaiian monk seals and other sea creatures.

And it's not just marine animals but the welfare of farm animals within the Meatrix that should be a significant concern for anyone who cares about animals. Even when using the most humane methods possible, farms are places of horror for animals. For example, in 2019, after an undercover investigation of Fair Oaks Farm in Indiana, touted as the "Disneyland of agricultural tourism," the Animal Recovery Mission released a video showing employees punching, kicking, throwing, and slamming baby calves and beating them with steel rebar.

A few days later, a CBS news article reported, "Footage shows Fair Oaks Farms workers dragging calves by their ears, throwing them into small plastic enclosures and hitting them with milk bottles." The article also stated that "calves appeared to stay in filthy, overcrowded and hot conditions. Temperature readings show it was more than 100 degrees Fahrenheit inside their hutches. Dead calves were dumped in mass grave sites by employees, the video shows."[243]

If these abuses occur at the "Disneyland" of farms with tourists mingling about, imagine what's happening at other farms where no one's watching. When animals are being exploited and treated as commodities, abuses occur. Period. There are plenty more undercover investigations that show similar widespread abuses and terrors that farm animals endure daily.

The Meatrix knows there would be a tremendous backlash if consumers understood how much suffering and abuse goes into the chicken leg or steak on their plate. Avoiding this backlash is why they lobby so hard to create the previously mentioned ag-gag laws to criminalize undercover whistle-blowing activities.

Speaking of suffering, male baby chicks are waste products in the egg industry. There's simply no way to make money on these innocent beings, so what happens to them? A person known in the industry as a "sexer" determines the sex of every chick born and separates the male chicks from the female chicks. Once separated, the female chicks become commodities of the Meatrix. The Meatrix disposes of the males in one of two ways: they are either ground up (eviscerated) alive or gassed—both of which the industry considers humane treatment.

As many of us were taught in school, nature seeks balance and will produce roughly equal numbers of males and females. Unfortunately, because of nature's balance the Meatrix kills approximately one male chick within days of being hatched for every female layer chicken in the egg industry. So even if you buy eggs that are from cage-free, free-range, humanely treated chickens, you're still supporting the deaths of billions of male chicks. "[Consumers are] funding this whenever they buy eggs. It's something that's inescapable no matter what label of egg you buy," says Emma Hurst from Animal Liberation, NSW.[244]

Generally speaking, the Meatrix treats animals as if their lives don't matter. As a result, farm animals are held captive, exploited, mutilated while fully conscious, forcibly impregnated, force-fed, and then starved before being "humanely" killed. Of course, these are "normal" farming practices and don't include any other abuse that may happen along the way, like those documented at Fair Oaks Farms.

Why does the Meatrix starve animals? I discuss many of these practices in Chapter 8, but it's standard practice to withhold food from animals at least forty-eight hours before slaughter. Food deprivation reduces traces of food in their digestive systems that would complicate the butchering process.

Even before the slaughtering begins, farmed fish—like land animals—are denied food for days before being killed just to keep the slaughterhouse cleaner. According to PETA, the Meatrix starves salmon for ten days before slaughter. Some might argue that this isn't cruel because salmon will not eat on their upstream journey to spawn. But that is a natural response to an innate drive, one they will never experience from living within the Meatrix.[245]

The oldest, most common, and inhumane way of slaughtering fish is asphyxiation. This method is cruel because fish can remain conscious (gasping for air) once removed from water anywhere from fifty to two hundred fifty minutes, depending on the species.[246] Another common way of slaughtering fish is an ice bath. In 1996, the Farm Animal Welfare Council said, "The cooling of live trout on ice after they have been removed from water should be prohibited."[247] But the practice continues.

A 2010 article published in *The Guardian* ends with the following observation: "We need to learn how to capture and kill wild fish humanely—or, if that is not possible, to find less cruel and more sustainable alternatives to eating them."[248] Some ethicists believe there is simply no humane method of animal slaughter.[249] Rather than spending valuable resources attempting to make an inherently cruel practice less so, I believe we should find alternatives to eating fish. Living outside the Meatrix is the only way to avoid contributing to the needless suffering of trillions of sentient aquatic creatures that simply want to be left alone to live out their lives.

There was a time when people said it didn't matter how fish died because they didn't have a sensory system or that other farmed animals lacked intelligence or feelings. However, science is showing us how wrong this old way of thinking is. For instance, one quick example is zebrafish. Zebrafish are the aquatic equivalent of mice, yet scientists have found them to be even more useful than mice for research purposes.[250] Recent data (April 2021) on zebrafish proves they have much more in common with us than previously thought: like humans, zebrafish lose their memory as they age, can get addicted to cocaine (as well as other opiates, stimulants, alcohol, and nicotine), remember their friends, are impatient, and feel pain.[251]

We kill land animals by the billions and aquatic creatures by the trillions each year, forgetting that each one is unique, with a life of its own—one it wishes to live out in peace. We know that animals—land, sea, and air—suffer, experience fear, and certainly do not wish to die. If we needed to eat animals to be healthy, that would be one thing. Still, I believe (and hope you will agree) that when there are healthier options for both ourselves and the planet, killing them simply for the fleeting pleasure it gives our palate cannot be justified. It is simply wrong.

The Meatrix doesn't want us to think about the lives of the animals we eat, let alone extend to them the same consideration we do non-farm animals. An example of this is that the US Animal Welfare Act, signed into law in 1966, which set minimum standards of care for animals used in research, bred for commercial use or exhibition—explicitly excludes farm animals from protection.[252] This omission is intentional. The Meatrix wants us to believe that farm animals are somehow different from laboratory animals and pets, and therefore, not worthy of being treated as anything other than commodities. I discuss this mentality in greater detail in the next chapter on speciesism.

Thankfully, Americans are increasingly becoming aware of animal welfare.[253] In 2010, then-President Barack Obama signed a bill, the Animal Crush Video Prohibition Act, that criminalized the creation of "crush" videos, in which animals are crushed, burned, drowned, suffocated, or impaled by humans.[254] Nine years later, the Preventing Animal Cruelty and Torture (PACT) Act passed unanimously in both the House and Senate and went even further to make these acts of torture, not just the creation of the videos, federal felonies.

This bill, written by a Democrat and Republican in a highly partisan climate, is a powerful testament to the universal, innate belief that animal cruelty is wrong. So naturally, law enforcement and animal rights groups applauded the signing of this bill. But like all other laws enacted to prevent animal cruelty, the PACT Act explicitly excludes "the slaughter of animals for food;

hunting, trapping, fishing, a sporting activity not otherwise prohibited by Federal law, predator control, or pest control."[255]

In terms of sheer numbers, the Meatrix kills and exploits millions more animals than are protected by the PACT Act. So why work together, unanimously, I might add, to eliminate some acts of cruelty to animals, and not *all* acts of violence to animals? These exclusions speak to the unjust prejudice we have for animals we deem as food and are another reason the Meatrix works to enact ag-gag laws to criminalize whistle-blowing activities. If people saw the innate cruelty of animal agriculture, there would be overwhelming criticism of the industry, which would lead to reforms the Meatrix doesn't want.

The number of animal deaths the Meatrix causes each year is incalculable. As far as the Meatrix is concerned, animal lives matter so little that many animal deaths—due to overcrowding, disease, heart conditions, and transit to slaughterhouses or fish markets—go unreported. Although some websites provide helpful counters, the truth is that their numbers are mere estimates. For example, many limit their counts to land animals or statistics for the US alone. Because of this, it's challenging to be sure of the exact number killed by the Meatrix each year.

Still, animal rights organizations try to quantify the number of animals killed annually for food. According to 2018 data collected by the UN Food and Agricultural Organization, more than one million land animals are slaughtered every hour in the US alone. That's more than twenty-six million each day, or more than nine-and-a-half billion per year. These figures are only for land animals in the US and exclude animals used in dairy and egg production.[256] When one includes all land and sea animals, the number killed each year in the US is much higher. According to the 2022 US Animal Kill Clock, "More than 55 billion land and sea animals die annually to support the U.S. food supply."[257]

The above figures are for US land animals; what about the number of fish killed each year by the Meatrix? Globally, there are one to three trillion wild seafood deaths every year. This figure excludes the estimated 37 to 120 billion killed each year on fish farms.[258] It's also worth repeating that no welfare standards exist for the one to three trillion fish caught in the wild each year. How is it that the welfare of meat, dairy, and eggs is of such concern that people are willing pay more at a grocery or restaurant for food with a "trustworthy welfare certification,"[259] but we care so little for the well-being of fish that we don't even have guidelines for their humane treatment?

Thinking of the treatment of fish reminds me of something I witnessed several years ago while traveling along the Mekong River. At an outdoor roadside food market in Cambodia, vendors selling a wide variety of fruits, vegetables, and seafood lined both sides of a street closed to vehicular traffic except for the ever-present mopeds. I observed a plastic washtub full of fish (I'm not sure which species, but I believe they were catfish) for sale sitting on the ground next to a vendor. The fish were rather large, maybe a foot in length, and packed so tightly in the basin they had no room to move around. The water barely covered their dark bodies. From what I could gather in observing them, they were slowly suffocating due to what I imagined was a lack of oxygen available to them. These fish, unable to move, appeared to be continuously gasping for air while the vendor sat indifferent to their suffering.

As a world traveler, I try not to judge and work to respect the people and culture of the country I'm visiting. Additionally, I realized the vendors at these outdoor markets had access to neither ice nor refrigeration. I assume many fish buyers, too, cannot keep meat fresh for days at home. So I believe I understood the circumstances that contributed to this scenario. As much as I hated seeing living, breathing, feeling creatures treated so callously, and because I was a guest in their country, I did nothing. Looking back, though, I regret my inaction that day. I wonder if the vendors were similarly mistreating dogs, would I have been more willing to intervene?

It's difficult for me to comprehend large numbers, like the up to three trillion marine lives killed each year. So I ran a Google search to see if I could get some perspective on just how large the number one trillion is. The Endowment for Human Development's website was at the top of my Google search results, so I clicked it. I'm glad I did because it offers multiple ways of visualizing large numbers.

One is a stack of one-dollar bills. A dollar bill is 0.0034 inches thick. A pile of one hundred one-dollar bills is 0.34 inches. One thousand is 3.4 inches. And one trillion one-dollar bills stacked on top of one another would be 67,866 miles high. To get a perspective of the estimated number of wild fish and marine animals killed each year by the Meatrix, a stack of three trillion one-dollar bills would go almost three-quarters of the way from the Earth to the moon.

Another way to imagine a large number is through a shopping spree. Let's say a person must spend twenty dollars per second, twenty-four hours a day, until they run out of money. For instance, a person with $100 to spend could only shop for five seconds, whereas a person with $1,000 could shop for fifty seconds. It would take 1,585 years for a shopper to spend one trillion dollars and 4,744 years to spend three trillion dollars!

Sadly, most people trapped inside the Meatrix are oblivious to their participation in animal cruelty, particularly concerning marine life. Not long ago, I had a conversation with a family member about their recent trip to New Orleans. They stayed with relatives and explained what a handful their relative's new dog was turning out to be. They told me this was their third pup, so I asked why they chose to get a third dog. The reply was that the family member was a "softie animal lover" and didn't want the stray dog to be in an abusive situation, so they adopted it. Soon after that—oblivious to the irony—I was told their "softie animal lover" host treated them to a delicious crab boil—using living crabs stored in a cooler until they were ready to be taken out and boiled alive.

It's difficult for me to fathom how a person can be an animal lover in one moment and an unthinking animal killer in another. But I know that this repeatedly happens because the Meatrix has taught us that only particular animal lives matter. Society teaches us to love dogs, cats, hamsters, dolphins, and whales and forcibly breed, exploit, torture, and slaughter cows, pigs, chickens, and fish. We study and go to great lengths to protect some species, such as the Hawaiian monk seal mentioned at the beginning of this chapter, while being utterly indifferent to the suffering of other sea creatures (that might not be as cute). Most people fail to understand that the line between lovable pets and all other living creatures, including farm animals, is arbitrary.

And this arbitrary line can easily be blurred if not erased altogether. A perfect example is Esther the Wonder Pig. Esther was billed as a "micro pig" in 2012 when Steve Jenkins and Derek Walter rescued and adopted her. Two years later, Esther weighed in at over six hundred pounds! Esther's great intelligence and sweet affection didn't go unnoticed, and she quickly became a member of the family. Steve and Derek write on Esther's website, "Esther is special to us, but she's exactly the same as all of her brothers and sisters who are not so lucky to be loved. All pigs are loving, intelligent and compassionate animals and they deserve better than the brutal life they are born into. Please consider leading a cruelty-free and compassionate lifestyle for yourself, and for all the Esthers out there in the world."[260]

Traditional media outlets like *HuffPost*, the *Washington Post*, *The Guardian*, and vegan news media such as One Green Planet,

PETA, and VegNews reported on Esther, and in two short years, Esther amassed hundreds of thousands of friends from all over the world. Then, in 2016 Steve Jenkins and Derek Walter published *The New York Times* bestselling book *Esther the Wonder Pig: Changing the World One Heart at a Time.*

Another story about farm animals being no different than dogs and cats is a more personal one. A dear friend of ours lives in the UK in an area that has wild chickens. A flock of wild hens migrated up the hill to her garden and hedges, and before long, a rooster followed.

Over time, the wild hens died of natural causes until only the rooster remained. When our friend's husband passed away, she developed a close bond with the lonely rooster and named him Big Boy. They kept each other company on the country estate, with her watching over, feeding, and getting veterinary care for Big Boy as he happily lived close to the house in her garden.

When Big Boy was far older than most roosters (especially wild roosters) can dream of being, this wild boy got sick for the last time. But because of the special relationship he had with his human friend, he happily stayed in her warm kitchen and enclosed porch, allowing her to feed him, medicate him, and bandage his legs. Our friend said her unique relationship with a wild rooster taught her that roosters really aren't different from dogs and cats—they're worthy of our love and affection and capable of showing appreciation and devotion in return. Our friend says she will never look at another chicken the same way again.

We've seen how our elected representatives repeatedly and intentionally exclude farm animals from legislation designed to prevent animal cruelty and that undercover reporting exposes even the best-of-the-best farms as places of abuse and horror for animals. We've learned about the mind-boggling numbers of land animals and marine lives killed by the Meatrix each year.

For these and other reasons, we owe it to the animals to escape the Meatrix.

HUMANS

If you remain unconvinced by the information I've presented so far and you still believe your health, the environment, or an animal's life is worth less than the pleasure you get from eating them, how about a human life? Would you agree that saving a human life is more important than your tastebuds? I hope your answer is yes.

> **Warning:** The following contains some disturbing information about child mortality rates.

In June 2013, *The Lancet*, a weekly peer-reviewed medical journal, published an article titled "Maternal and child undernutrition and overweight in low-income and middle-income countries." The report states that there were 3.1 million nutrition deficiency-related deaths in children five years of age or younger each year globally. This rate means that as of 2013, 8,493 children died each day from malnutrition—more than one every ten seconds.[261]

According to Vandana Shiva in *Stolen Harvest*, "Overall, animal farms use nearly 40 percent of the world's total grain production. In the United States, nearly 70 percent of grain production is fed to livestock."[262]

This diversion of food from humans to animals isn't sustainable. It contributes to a health crisis in most industrialized nations and helps fuel global climate change, mass extinctions, biodiversity loss, and pandemics. Plus, it's a contributing factor in the malnutrition deaths of 3.1 million children aged five years or younger each year worldwide. The benefit of a plant-based lifestyle to humanity becomes abundantly clear when you consider these 3.1 million malnourished children and add the 8.1 million human lives Oxford University estimated would be saved annually by the year 2050 if everyone adopted a plant-based lifestyle. So if you won't go plant-based for your health, the environment, or the animals, please keep the humans in mind.

SPIRITUALITY

For most of my adult life, I've believed that organized religion holds no more of a monopoly on spirituality than the Meatrix does on protein and calcium. One needn't participate in organized religion to live a spiritual life. To me, being spiritual means striving to do as little harm as possible, reducing suffering where I can, and making a positive difference in the world. I believe living a spiritual life means speaking out against injustice and embracing an ever-growing circle of compassion. I try to be forgiving and extend love to others daily. All of this is easy to say but sometimes hard to do.

Using this broad definition of spirituality, I hope that most of us would agree that we're trying to be better, more spiritually-minded people. Yet as I write this, I realize that I didn't spontaneously embark on my plant-based journey in a vacuum. Instead, my spiritual mentor encouraged me to do so, and I received support from my chosen family and friends. No one is an island, and spirituality for me includes accepting our interconnectedness and interdependence.

Americans pride themselves on their independence and individualism, yet observing the spread and outbreaks of the COVID-19 pandemic has, among other things, been a powerful reminder to me of just how interconnected and interdependent the world is.

Speaking of interconnectedness reminds me of a quote I recently read attributed to Chief Si'ahl (namesake of the city Seattle): "Humankind has not woven the web of life. We are but one thread within it. Whatever we do to the web, we do to ourselves. All things are bound together. All things connect."

The realization of the interconnectedness of all life helps guide me to make what I consider to be better choices. And it's not just me. For example, when renowned vegan physician Dr. Michael Klaper adopted plant-based eating, he made different choices, which I attribute to acting on spiritual principles.

Dr. Michael Klaper's story began with a single comment from a friend. Dr. Klaper spent the early part of his medical career working as an anesthesiologist in the emergency room of a Chicago-area hospital. Witnessing the daily gruesomeness of violence changed him profoundly, and as a result, he dedicated himself to living a life of nonviolence. He was sharing his newfound commitment with a friend during a dinner out one evening, and near the end of the meal, his friend said this was all well and good, but "while you're looking around for places in your life to eliminate violence you might start by looking at the piece of meat on your plate and what it took to get there." Dr. Klaper said when he paid for his steak dinner that night, he felt complicit in the cow's death, and he has been plant-based ever since.[263]

I mention Dr. Klaper's story because we never know the impact we might have on another person. I'm including it here because it feels very grounded in spirituality and the interconnectedness of all things.

Maybe as you're reading these pages, you're beginning to feel an urge to make different choices—choices that will benefit your health, the environment, and the animals. I encourage you to pay attention to that inner urge and follow your heart.

Speaking of following your heart, many young children seem to be naturally empathetic toward animals. There are videos of children online (on YouTube, search "vegetarian children") who refuse to eat their chicken nuggets when they discover their origins or give up eating bacon after watching the movie *Babe*. I see this innate tendency toward empathy as one of the spiritual gifts we were all born with. Fortunately, because of the compassion children often feel, they conclude it's wrong to kill animals just for food.

What comes of these children and their newfound awareness of the origins of the food on their plate? Some will continue to forgo eating chicken nuggets and bacon, but for most, their refusal to eat animal products becomes no more than a "phase." Maybe they lack parental support or simply decide the pleasure they get from eating the nuggets or bacon outweighs the harm done to some abstract, faceless, nameless chicken or pig. But I bet it's primarily due to the societal pressures the Meatrix places on all of us to ignore our inner empathetic, spiritually interconnected feelings. When these children give up their nuggets, they're taking the red pill, but most are lulled back into the clutches of the Meatrix at the unfortunate price of turning off one of their spiritual gifts: the gift of empathy for all living creatures.

You may be thinking it's wrong for me to generalize about all children based on a few YouTube videos, and you'd be correct. Fortunately, a 2021 study from the *Journal of Environmental Psychology* revealed that 84 percent of children believed it was wrong to eat cows, and 79 percent said it was wrong to eat pigs. The same study showed that children were amazingly unaware of the origins of the food on their plates and often mistook animal-based foods as coming from plants. For example, 41 percent of children said bacon was a plant-based food. The report said the children's lack of understanding of food production was a result of little to no exposure to farms, not being taught in school where food came from, and parents who, because of their own conflicted feelings about eating meat, shielded their children from the realities of food production. The authors said that "some parents instead skirt the truth altogether through vague terminology that has potentially lasting impacts on children's eating habits."[264] Scientists in the study further believe that being more transparent with children about food production might be a way to help mitigate climate change. This is how the "tradition" of eating meat is passed on from one generation to the next. Through deceit and disassociation, children often become unsuspecting meat eaters, indoctrinated into the Meatrix without their consent.

Furthermore, in April 2022, a study was released saying, "Compared with young adults and adults, children (a) show less speciesism, (b) are less likely to categorize farm animals as food than pets, (c) think farm animals ought to be treated better, and (d) deem eating meat and animal products to be less morally acceptable."[265] From a very early age, children are concerned with moral concepts and avoiding harming animals. It appears the transition away from such thinking begins to happen in late childhood and early adolescence when the moral acrobatics required for eating meat are established. The study ends by saying, "Human food production and consumption are related to timely global issues like climate change. Attempts to mitigate these global problems might benefit from open dialogues regarding our relationships with animals. The evidence presented here suggests these dialogues ought to begin in youth when the social construction of the way humans think about animals begins."[266]

Thinking of naturally empathetic children reminds me of the Golden Rule. As a little boy at the Good Shepherd Catholic School, I learned the Golden Rule (do unto others as you would have others do unto you). So naturally, because it was taught at my parochial school, I assumed it to be one of Christ's teachings, but I have learned it precedes Christ's life by centuries.

Harry J. Gensler, the author of the book *Ethics and the Golden Rule*, has created an impressive chronology of the Golden Rule. According to Gensler, perhaps the first written example of the Golden Rule is ancient Egypt's "Eloquent Peasant" story from 1800 BCE.

While the Golden Rule is part of the moral code of many of the world's religions, you don't need to be religious to endorse it. The rule is all about our actions toward others. For example, I want others to treat me with love and respect, so I try to love and respect others.

What if we treated all living creatures as we would like to be treated? Would we arbitrarily harm others when it's not necessary? Would we needlessly inflict pain and suffering onto others? This book shows that the harm humans do to animals *is* unnecessary. Science shows us animals are complex beings with individual personalities, greater mental capacities than we previously understood, and greater emotional intelligence than we typically give them credit for. And perhaps most importantly, like all animals, including humans, they don't want to die! For this reason, I think escaping the Meatrix and becoming plant-based is simply living the Golden Rule of treating others as you wish others to treat you.

My desire to live a spiritual life, along with many other reasons listed in this book, motivates me—and an exponentially growing number of others—to eat only plants. However, we all have free will, and most of us have choices as to what we eat. For me, my spiritual path directs me to make my choices based on love.

We know there are consequences every time we consume an animal product. We also are aware of the pain and suffering our choosing animal-based foods causes trillions of helpless beings. And we're also now aware of the environmental destruction and accelerated climate changes associated with the Meatrix. For these reasons and more, I hope spiritually-minded people will want to make better choices such as not funding animal cruelty, starting with the food on their plates. I hope we will all make choices based on the Golden Rule of empathy and, above all, love—for ourselves, others, and our planet.

FASCINATING FACTS

- US Animal Welfare Act of 1966 excludes farm animals from protection.
- An estimated 55 billion land and sea animals are killed in the US alone for food each year.
- An estimated 1–3 trillion aquatic deaths are attributed to the Meatrix annually.
- 70 percent of US grain is fed to livestock annually.
- 84 percent of US children four to eight years of age say it's wrong to eat a cow.
- 41 percent of US children four to eight years of age think bacon comes from plants.

CHAPTER 6

Speciesism

Both my parents grew up in downtown Louisville, Kentucky. While my father had a dog and a bird as pets growing up, my mother had none. She spent her summers on her grandparents' farm and believed that animals belong outside, on the farm. As a result, I grew up in a home without pets.

As an adult, I came into a household with a dog and two cats, so I learned to not only step around litter boxes but clean them out—to not only walk the dog but pick up dog poop. It was immersive therapy that, looking back, I was ill-equipped to handle. But kitties purring on my lap and a sweet dog just wanting to love can change a mind and heart pretty quickly. Fast forward to today, where I have two rescued pups that cuddle with me, play with me, and love me, and whom I love as the family members they are.

I realize that not everyone shares my beliefs about the ability of pets to become family members. Still, I hope this book, and especially this chapter, will help you see animals and our relationship with them in a new light.

Speciesism, the idea that some animals' lives are more important than others, is the root of the Meatrix. But, speciesism, like all other "-isms," is not innate, but something we learn. It's a human construct that research tells us doesn't coalesce in children until late childhood or early adolescence.[267] Speciesism teaches us that a life form's position in the hierarchy can determine its value and that humans have control over all the less-fortunate non-human animals below them. Unfortunately, this type of thinking has some profound and dangerous consequences.

One is the immeasurable suffering that zoonotic pathogens have unleashed on humanity over the past ten thousand years, which was discussed in Chapter 3. Another is the vast impact animal agriculture has on climate change—it's responsible for up to 51 percent of all greenhouse gases, as mentioned in Chapter 4. Both are largely because we encroach into animal habitats, hunt and kill them, and steal their calves to drink their milk. In past centuries, we also used them to pull our carts, plow our fields, and provide our transportation.

The arrogance of speciesism fuels the belief that nonhuman species deserve fewer rights and protections because they are less intelligent, lack emotional capacity, and aren't self-aware—and, therefore, are less important than humans. However, the assumption underlying this thinking is flawed. Science is proving that even under the immense stress of farming, animals form bonds with each other, express strong emotions, and exhibit intelligence in ways we're just learning to recognize.

For instance, studies have shown that pigs can identify objects, have long-term memory, and can be aware of time differences. And anyone who has had a pet pig would tell you they're affectionate, easily housebroken, and, like dolphins and sea lions, can learn any trick a dog can, such as fetch a Frisbee.[268] That pigs can do any of the above is astonishing when one considers that, unlike dogs who have been selectively bred for nearly ten thousand years to pull sleds as well as become "man's best friend," pigs have only been selectively bred to become as big and as profitable to the Meatrix as possible.[269] How much smarter and companion-like would pigs become if we selectively bred them to the same extent we have canines to become our loyal and faithful companions?

An article by neuroscientist and expert in animal behavior and intelligence Lori Marino explains how chickens demonstrate self-control by foregoing immediate gratification for a later greater reward—something that doesn't consistently emerge in humans until around the age of four when the mental processes that allow for self-awareness begin to develop.[270] And cows express positive emotions and excitement when learning a new task.[271] So, imagine how happy some cows in New Zealand must have been upon learning the new task of being potty trained quicker and easier than human toddlers. You might wonder why anyone would potty train cattle, but it turns out New Zealand and Australia graze nearly all their cattle on open pastures, and their nitrogen-rich urine contaminates the soil and nearby streams, lakes, and aquifers, which can create algae blooms that can be deadly to fish and other aquatic creatures.[272]

Nitrous oxide, a long-lasting greenhouse gas that occurs when a cow's nitrogen-rich urine breaks down in the soil, is three hundred times more potent than carbon dioxide at trapping heat in our atmosphere. Nitrous oxide accounts for approximately 12 percent of New Zealand's greenhouse gas emissions, with most of it coming from agriculture.[273]

Potty training cattle might be a way to capture and treat a cow's urine, preventing it from contaminating soil and watersheds as it currently does. If a cow can be potty trained more quickly than human toddlers, how can their intelligence be denied? This story is another example of cattle being more than food commodities, and highlights just one of the horrible environmental impacts of animal agriculture.

In 2018, *Smithsonian Magazine* shared some of the most recent discoveries about animal grief, saying it's not a question of *whether* animals grieve, but *how* they grieve. The author believes we know so little about animal grief simply because scientists haven't bothered to look. Clearly, our lack of awareness or curiosity about a thing does not mean it doesn't exist.[274]

Research done by Krista McLennan of North Hampton University reported the benefits of long-term social connections in cows. Unlike an unfamiliar cow, cows paired with a friend exhibited lower heart rates and less overall stress.[275] Findings of a 2014 University of British Columbia study demonstrated that cows are not only calmer when they're with a friend; they're smarter too. The study showed that cows that grew up with a buddy learned more quickly. Furthermore, social housing "may result in animals that are more flexible in their responses to changes in management and housing."[276]

Humans have long believed that their language development and understanding of numbers sets them apart from all other animals. But science reveals the fallacy of that line of thinking. Species such as birds and bees, whose evolution diverged from mammals hundreds of millions of years ago and who therefore have much smaller and differently wired brains than ours, can understand numbers.[277] In fact, in some cases, they can even comprehend the mathematical concept of zero—a concept so abstract most children are unable to grasp it until age six.[278]

Trained chimpanzees who have learned to associate numbers with symbols can put those symbols in ascending order. This achievement might not surprise many people since chimpanzees and bonobos are our closest living relatives in the animal kingdom, sharing an incredible 98.8 percent of their DNA with us. But scientists have recently discovered that lions, ants, spiders, bees, crows, frogs, and domesticated chicks can also understand numbers.[279]

A 2009 article reported on an experiment in which days-old baby chicks were allowed to imprint with five objects. After several days, the five imprinted objects were placed behind two screens (three objects were behind one screen and two behind the other). The chicks always chose the screen with the largest number of imprinted objects behind it. After the scientists visibly transferred objects, one by one, from behind one screen to the other, the chicks again chose the screen with the largest number of imprinted objects behind it. The experiment shows the numerical intelligence of days-old, domesticated baby chicks. The authors wrote, "Results suggest impressive proto-arithmetic capacities in the young and relatively inexperienced chicks of this precocial species."[280]

And the intelligence we're beginning to recognize in farm animals isn't limited to land creatures. With their lack of cuddly fur and the presence of scales, most people find it difficult to relate to fish. But this might start to change as new data about fish emerges. For instance, did you know fish enjoy physical contact? Similar to cats who enjoy rubbing against our legs, fish enjoy gently rubbing against one another. Other recent studies have revealed that fish are aware of time and enjoy playing, and some, like goldfish, are studied because of their capacity to learn and their impressive memories. In fact, Professor Felicity Huntingford of the University of Glasgow, who has spent more than fifty years studying fish, says goldfish have been studied as far back as 1908 because scientists recognize goldfish learn tasks similar to mammals and birds.[281]

These examples represent only a few of the many recent scientific studies regarding farmed animals' social and mental intelligence and self-awareness. Scientists have long studied primates and other mammals, but the fact that psychologists and animal behaviorists are now researching and publishing data regarding farmed animals indicates that humans are beginning to recognize they have been seeing it wrong this entire time.

Our newfound interest in farmed animals' lives and the studies done thus far reveal that the experiences of these creatures are much richer and more complex than we ever imagined possible. Of course, the means to study and observe farmed animal behaviors have always been available to us, but we've simply avoided the topic until recently.

Even if, despite the science to the contrary, one argues that food animals lack intelligence, aren't self-aware, and don't have friends—and are therefore unworthy of our consideration— remember there are, of course, humans who are also *not* intelligent, lack self-awareness, and are unable to sustain friendships. Additionally, some people are born with limited capacities for learning or never attain the social skills to create friendships; they could have medical problems that make them unable to think well. Should their rights be revoked because they've lost, don't have, or never will achieve a level of intelligence, self-awareness, or social skills similar to the rest of us? Certainly not. Why can't we use the same logic when it comes to farmed animals?

You might ask if I'm saying that an animal's life is more important than or as important as a human's life. Of course, most people would consider a human life inherently more important if forced to choose between the two. However, the operative words are "forced to choose." Fortunately, we don't need to take the life of an animal for food.

I think animal rights activist James Aspey summed it up best. When a skeptical interviewer seemingly tried to trip him up by asking if he was saying a squirrel's life was more important than a human life's, he replied, "I think your life is more important to you, and the squirrel's life is more important to it."

It's easy to understand how an individual squirrel, or any individual animal, would place more value on its own life than yours. The instinctual drive to survive is strong within every unique creature. But we refuse to see animals, especially farmed animals, as individuals. Instead, we choose to see them as herds, flocks, schools, or commodities rather than groups of unique beings.

In speciesism, humans de-individualize the species they oppress, strip them of their uniqueness, and refuse to see them as complex, nuanced individuals. When entire groups are "othered" like this, it fuels greater and greater oppression.

In her article on The Swaddle, culture editor Rajvi Desai explains:

> When we understand the "others" only on the basis of their group attributes—ethnicity, skin tone, class, religion, etc.— we fail to understand their lived experience, making empathy impossible. Thus, we fail to appreciate their suffering, understanding the suffering of our own and adjacent groups as tragic, while dismissing theirs with a mere shrug.[282]

I believe that the "othering" of farmed animals allows most people to dismiss their suffering just as callously. Few people are capable of shrugging off the suffering they see of dogs and cats. Most will try to help a dog that's been hit by a car or a skinny kitten meowing at their back door, and some will rescue them and adopt them as pets. However, few are willing to bridge the "othering" gap and extend the same empathy and care to farmed animals. Why is that?

Empathy requires emotional and mental labor, an activity many people often have difficulty doing for humans who appear distinctly different from themselves, let alone for animals who are an entirely different species. Compounding the lack of empathy for farmed animals by those within the Meatrix is that these individuals benefit directly (but not in the long run) and ultimately derive pleasure from eating animals. It makes sense they would avoid empathy's emotional and mental labor by giving little consideration to these animals. Not considering the plight of farmed animals is also an example of a dissonance-reducing strategy (see Chapter 10).

It's important that we see animals as individuals because we often have more empathy for individuals than large groups, whether they are humans or animals. For example, in the US, many people see dogs and cats as unique individuals. We're able to do this because many of us spend time with them, get to know them, invite them into our homes, recognize their intelligence, and appreciate the quirks of their personalities. Because we see these creatures as unique individuals, we empathize with them, feed and shelter them, nurture them back to health when they are ill, love them, and mourn them when they die. Many people are more than willing to do these emotional and mental labors for their pets.

For example, I get immense pleasure out of sharing my life with our two pups. They're littermates and have a lot in common, but I appreciate their distinct personalities. Cullen is laid back and enjoys sniffing around the house and yard. Finn is more athletic and loves chasing after tennis balls. In addition, Cullen is a more cautious introvert, whereas Finn is a trusting social butterfly.

There are stories of cows, chickens, and pigs who befriend humans when given a chance and vice versa. Prime examples of this are Esther, the pig, and Big Boy, the rooster, mentioned earlier in this book. It's obvious to those who befriend farm animals that they are receiving love and affection in return. In addition, due to their bonding, these people are much more willing to do emotional and mental labor for these creatures.

I believe most people tolerate the ill treatment of farmed animals due to biases the Meatrix has taught us throughout our lives. When we have negative biases toward farmed animals, the Meatrix wins. Maybe if chickens looked like dogs and fish looked like cats, we wouldn't have the same biases and would be less likely to "other" them and then butcher and eat them.

If we can escape the Meatrix and thrive without it, why do we kill trillions of land and sea creatures every year?

FASCINATING FACTS

- Nitrous oxide, a long-lasting greenhouse gas that occurs when a cow's nitrogen-rich urine breaks down in the soil, is 300 times more potent than carbon dioxide at trapping heat in our atmosphere.
- Crows can comprehend the mathematical concept of zero—a concept so abstract most children are unable to grasp it until age six.
- Days-old domestic baby chicks have impressive proto-arithmetic capacities.

The Cult of the Meatrix

Your brain is hooked on the shit that the Matrix
has been force-feeding you for years.
—Morpheus to Neo, *The Matrix Resurrections*, 2021

Like Neo in the Matrix, the Meatrix hooked my brain and my body on what it force-fed me since birth. Since taking the red pill and becoming plant-based, I see how I was living in a world of speciesism, brainwashed by the Meatrix and hooked on its products. I believe the Meatrix held this power over me because it is a cult I was born into and never questioned.

Robert Jay Lifton, a psychiatrist who once taught at the Harvard School of Medicine, wrote the seminal paper on cults in 1991. According to Lifton, there are three primary characteristics shared by a destructive cult:

1) a charismatic leader who increasingly becomes an object of worship as the general principles that may have originally

sustained the group lose their power; 2) a process I call coercive persuasion or thought reform; 3) economic, sexual, and other exploitation of group members by the leader and the ruling coterie.[283]

Regarding the first characteristic, it's accurate that humans need to consume protein, but the Meatrix has replaced the general idea with the specific one that we need *animal* protein. The perceived need for animal protein is a classic example of a principle that initially sustained a group (the need for protein) losing power to a charismatic leader (the Meatrix) that becomes an object of worship, replacing the initial idea that was sustaining the group.

Many people insist meat be a part of every meal and staunchly defend every bite they take while deriding plant-based proteins and those who eat them as weak. This disdain for plant protein is a perfect example of how the Meatrix has brainwashed us to worship meat as though it was a powerful cult leader.

For example, I absorbed the idea of meat being good and going meatless being bad at a very early age. I can vividly and fondly recall being a child sitting with my siblings at our kitchen table. We were eating our cereal while my mom cooked breakfast for my father before he headed off to work. Every day he ate bacon, two eggs fried over easy, toast, and coffee. As my mother cleaned up, she carefully poured the bacon grease from the skillet into a metal container she kept under the kitchen sink. She used this grease to fry eggs and other foods such as potatoes, or she added a dollop into the pot with green beans. Using bacon fat was "Southern cooking," and it was something my mother learned from her mother.

But even before my mom had finished cleaning up the breakfast dishes, she would ask, "What are we going to have for dinner tonight?" She wasn't asking for our input as much as she was thinking aloud about what meat she would need to remove from the freezer to thaw in time for the evening meal. Once she selected the meat, she determined the accompanying vegetables. Still, the side dishes played second fiddle to the meal's centerpiece: the meat. As Catholics who observed Lent, we ate fish rather than go meatless on Fridays. At home, meat, whether from the land, air, or sea, was always the first consideration in determining a menu.

My elementary schoolmates and I resented the hamburgers served in the cafeteria at the small Catholic school we attended. To feed more children, the chef extended the beef with oats. My friends and I believed these burgers, with their added plant-based ingredients, were an inferior product. We felt as though the cafeteria was slighting us. At that early age, we had already internalized the message that it was better to have 100 percent all-beef patties!

This feeling of being entitled to meat three times a day is the America of my youth—one in which almighty meat, the cult leader of the Meatrix, was the centerpiece of the plate. We weren't happy if meat wasn't part of every meal—or, like my grade-school burgers, if it wasn't 100 percent pure meat.

This same mentality was reinforced in public high school by my new friends. The cafeteria had a traditional food line but also offered a burger-and-fries station. Of course, being teenagers, we often gravitated toward greasy burgers and fries. Even when choosing those burgers over other food offered, we complained about the patty's quality, referring to it as "mystery meat." We were suspicious that they might have sneaked some soy protein or, heaven forbid, oats into those patties!

Nowadays, when grilling up a plant-based burger, I think of the 180-degree turn I've made. Going from being offended at not having 100 percent beef patties in school to happily and purposefully choosing 100 percent non-meat patties today makes me chuckle.

I think the overwhelmingly cultlike appeal, blind following, and power that meat has over most of us is due to advertising. In the 2019 documentary *Game Changers*, which dispels the myth that optimum athletic performance requires animal protein, actor, producer, bodybuilder, and former governor of California Arnold Schwarzenegger says the prominence our society gives meat is due to clever marketing.[284]

An example of the clever marketing that Arnold Schwarzenegger referred to began in 1992, when the Beef Industry Council launched an ad campaign that initially ran for seventeen months at $42 million. The campaign slogan: "Beef. It's what's for dinner." This beef campaign slogan was so successful that almost thirty years later it is still recognizable by more than 88 percent of Americans.[285]

Another ad campaign, this time by the fast-food franchise Wendy's, used the catchphrase, "Where's the beef?" This memorable slogan was a stab at Wendy's competitors' smaller, frozen patties. The phrase became so popular that it was transformed from a question about meat to a phrase that generally questioned the substance of *any* idea or product. "Where's the beef?" was cemented in the minds of Americans, reinforcing the idea that a meal lacks substance without meat at the center of it.

Meat itself has become an object of worship. The wealthy enjoy eating it in abundance regularly, and the poor wish to do so. It's embarrassing to admit, but I felt ashamed eating the burgers served to me in grade school. I believed extending the meat so it could feed more hungry children was a sign of our poverty. I had already internalized that eating meat at every meal was a way of telling ourselves and others we were not poor.

On the evening of my high school prom, I took my date to dinner. It was a special occasion, so, having been brainwashed into the cult of the Meatrix, I treated the two of us to a romantic dinner of steak and lobster. I had never eaten lobster before but knew, based on what I had observed through marketing, television, and movies, that a surf-and-turf meal is high class, and I wanted to impress my date. Moreover, it made me feel grown-up and "manly" that I could afford to provide such a meat-laden feast to my date.

Of course, meat was also the centerpiece of every holiday meal and celebration growing up. We had turkey on Thanksgiving, ham on Easter, and pot roast on Sundays. For Christmas, we often had both country ham *and* turkey! On July 4, we grilled hamburgers and hot dogs. Meat, meat, meat.

Nothing in recent memory epitomizes the charismatic, cult-like nature of the Meatrix to me more than the Arby's commercials with the slogan, "Arby's: we have the meats!" The catchphrase, spoken in the deep, bass voice of actor Ving Rhames, sounds both celebratory and proclamatory—Arby's doesn't just have meats, it has *the* meats. The only thing missing is the preceding fanfare of trumpets.

We have been relentlessly bombarded with message after message that we want and need the Meatrix's products throughout our lifetime. As a result, we've made powerful associations and hard-to-break habits that border on addiction not only when it comes to festive occasions but for every meal.

It's difficult to kick a habit when it's supported by the government, powerful industries with substantial advertising budgets, your friends and family, and the rest of society. There are many forces at work that intentionally strive to keep us addicted to meat. As a 2020 *Insider* article laments, "I would love to see the commercial farming industry dismantled. I would even be happy to participate in that dismantling. But as long as meat appears in front of me and I can afford it, I will eat it."[286] That sad statement epitomizes the addictive, charismatic allure of the Meatrix. Even when someone wants to kick the habit, they're often unable to do so.

We were born into the Meatrix, so from birth we have been brainwashed (Lifton's second characteristic) into thinking that meat eating is normal. The Meatrix tells us that humans are at the top of the food chain, that we deserve meat and have a right to kill animals to serve our desire for its meat (classic speciesism).

On the other hand, science tells us humans are not at the top of any food chain and that consuming animal products contributes to climate change and many of the common illnesses plaguing our society. As we've discussed, plant-based diets prevent certain types of cancers and are associated with lower rates of cardiovascular disease, type-2 diabetes, obesity, total cholesterol, and strokes.[287] And yet, we're so addicted to meat we'd rather destroy the planet and become ill than stop eating it!

The Meatrix has most people doing things that are not in their own best interest but consistently in the best interest of the Meatrix's leaders. Contrary to a mountain of evidence, the Meatrix has consumers engaged in activities that lead many of them toward chronic disease and an early grave, while the Meatrix reaps the profits.

The Meatrix would rather people eat meat three times a day than only on special occasions (Lifton's third criterion). For this reason, the Meatrix exploits the masses economically by keeping prices artificially low, allowing people to easily purchase their meat-based foods.

And while the Meatrix profits off the economic exploitation, it also exploits the health of group members. The Meatrix has convinced most people that we need meat for protein and dairy for strong bones, and without them we will suffer.

Similarly, in the movie, *The Matrix*, Neo is indoctrinated into thinking his life is normal. Without his growing awareness and curiosity, he might have spent his entire life in the clutches of the Matrix, never meeting Morpheus and taking the red pill.

Those who spend their entire lives living within the cult of Meatrix, like Neo, are unaware they're doing so. Through a lifetime of catchy advertising and social pressure, the Meatrix is brainwashing people so that consumption of animal products becomes as natural to us as the air we breathe. Breaking free from the clutches of the Meatrix cult isn't likely to happen without some sort of conscious disruption. It's precisely this type of intervention that I hope this book is providing.

One of the most powerful scenes in the movie occurs after Neo swallows the red pill and begins to free himself from the Matrix. Of course, his freedom is not instantaneous. In fact, the scene is quite dramatic as he struggles to physically disentangle himself from the Matrix and the thick, viscous goo in which he's spent his entire life.

Like Neo, freeing yourself from the cult of the Meatrix may not be immediate. It will likely have its challenges, as seemingly meat-free, plant-based products often contain animal products and byproducts. The best way to ensure you're entirely free from the Meatrix is to educate yourself about the many ways that the Meatrix incorporates hidden animal products into our food supply. Your education isn't going to happen overnight any more than Neo's newly awakened understanding of the intricacies of the Matrix did. I'm still learning some of the ways animal products are hidden in our foods. I recommend you educate yourself and give yourself a break if you unknowingly consume something not plant-based. If you're like most people in the United States and other Western countries, you've been plugged into the Meatrix and immersed in its goo your entire life. But like Neo, once you're free, you will find that your body is stronger and more capable than you previously imagined. And like Neo, you will see the world for what it truly is and will want to free others from the Meatrix's clutches.

Continuing with the idea of the Meatrix as a cult, Rick Ross, director of the Ross Institute for the Study of Destructive Cults, Controversial Groups and Movements, wrote an article listing ten warning signs that a group may be unsafe. On that list is "There is no legitimate reason to leave, former followers are always wrong in leaving, negative or even evil."[288]

We see an example of this line of thinking in *The Matrix*: before Neo met with Morpheus, Agent Smith tried to convince Neo that Morpheus was a terrorist. Smith attempted to recruit Neo to help the agents capture him. In your own life, you may have heard or read negative opinions from those within the Meatrix about plant-based eaters. Over time you may have adopted some of those beliefs for yourself. For example, veganism is extreme—it goes against dietary norms and is antisocial; vegans are unhealthy, physically weak, and not manly.

To prevent others from defecting, both the Matrix and Meatrix brand those who leave the "cult" as negative or evil. In *The Matrix*, Neo refuses to go along with Agent Smith's plan and meets with Morpheus, and when Neo is finally free, the Matrix fights hard to recruit him back. Of course, Neo refuses, which leads to some epic fight scenes between Neo and Agent Smith. When you decide to leave the Meatrix, you may encounter pushback from friends and family. But it's simply those within the Meatrix trying to recruit you back.

They may look at you like you're crazy for not eating a juicy steak or a bowl of ice cream. Sometimes there's pressure to come back to the Meatrix. For example, some people will try to lure you off your path by telling you how good their food tastes, that you're ruining a family holiday meal, or that you're harming your health or don't know what you're talking about when, in fact, you now know the science, and you do know what you're talking about—you've been there, done that, and read the research!

Generally speaking, human beings are herd animals who are afraid to leave the safety of the tribe to go into the woods alone. Members of the tribe will stoke your fear by saying you can't go into the woods alone because the bears will get you, or you can't give up animal products because you'll ruin your health or become antisocial. For this reason, I believe it's paramount that everyone willing to explore dietary change and escape the Meatrix find a plant-based group (online or in-person) for support. Associating with a like-minded group can make all the difference between feeling alone and realizing you're part of a growing movement that's helping to slow climate change, improve people's health, conserve natural resources, and spread love and compassion around the planet. If your family isn't on this journey with you, I highly recommend surrounding yourself with supportive friends. It can make a huge difference.

FASCINATING FACTS

- 88 percent of Americans still recognize the "Beef. It's What's for Dinner" campaign slogan almost thirty-three years later.
- The 2019 documentary *Game Changers* dispelled the myth that peak athletic performance requires animal protein.
- Cults often brand those who leave as wrong, negative, or evil.

CHAPTER 8

Isn't Veganism Extreme?

Warning: The following contains graphic information of routine practices within the Meatrix you may find disturbing, such as forced impregnation of dairy cows and chickens.

The Meatrix wants people to think that plant-based eaters are extreme or "extra." But to me, meat eaters, not vegans, are the extreme ones. Plant-based eating advocates a straightforward cycle of growing plants for human consumption. It's simple and would feed the world and benefit the earth and the health of all animals, including humans.

In previous chapters, I discussed the wasteful return on investment that animal agriculture is. But in this chapter, I want to focus on some of the truly "extreme" and cruel farming practices required for you to have a cold glass of cow's milk with your cookies or a cheesy slice of pizza.

We all know that for a mammal to produce milk, it needs to undergo the hormonal changes associated with pregnancy and birth. But few people stop and think about how dairy cows in the Meatrix become pregnant. In *The Matrix*, AI is an acronym for artificial intelligence. But in the Meatrix, AI stands for artificial insemination—a euphemism for forcible impregnation. Most farm animals in developed countries—and a growing number in developing countries—get their start in this world via artificial insemination, and dairy cows are no exception.

For cattle, the AI tale begins with a bull. A 2017 *VICE Magazine* article tells of Simon Amor, who has been in the business of collecting bull semen for twenty years. Yes. That's a job within the Meatrix. Simon says that after a bull is brought to a cow in heat to arouse it, it's allowed to mount the cow. But at that point, Simon's job is to quickly redirect the bull's penis into a heated artificial vagina, which has a vial attached to collect the semen. Like the cows at the dairy, the bulls never get to experience copulation the way nature intended. At the dairy in Melbourne where Simon works, they collect semen from twenty bulls a day, two days a week. They artificially collect the bull's semen from the time the bulls are eight months old. At that age, they are easier to handle and more easily trained to accept the process. The collectors work to keep the conditions very serene to prevent the young bulls from ejaculating on the floor, sending valuable sperm—and money—down the drain. One collector reported that it took a year to get used to the odor of bull semen, and he still gets squeamish at times.[289]

Bull semen is not only collected for local artificial insemination programs, but it gets exported around the world. The US and Canada have dominated the bull semen market for years. In 2020 the US accounted for 49 percent of the world's bull semen ($251 million), while Canada accounted for 16.7 percent ($85 million).[290]

Semen collection is only one part of the process. After that, it gets even more extreme.

Impregnating dairy cows in the Meatrix is the bovine equivalent of sexual assault. The first step requires immobilizing cows in what the industry actually calls a "rape rack." Then, to get the semen inside the cow, workers typically put their left arm deep inside the cow's rectum. In reaction to this assault, the cow's strong rectal muscles compress the reproductive tract back into the pelvic cavity. This compression creates folds within the cervix that then must be elongated with the left hand by grasping the cervix and gently moving it toward the cow's head. Once the left arm has fully elongated the cervix, then, using their right hand, they insert an insemination gun into the cow's vagina, through the cervix, and into the uterus, where the bull's semen is finally injected.

And people think plant-based eaters are extreme?

THE ARTIFICIAL INSEMINATION OF CATTLE

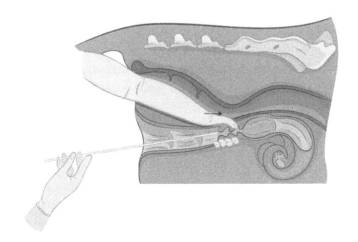

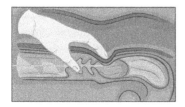 

InfoVets, "Reproduction Management: Artificial Insemination" (2005),
www.infovets.com/healthycowinfo/A716.htm

Fast forward a little more than nine months, and the new-born calf is separated from its nurturing mother within hours of being born.

This separation results in the disturbing sounds of cows and calves loudly and continuously calling out for each other. According to Newbury, Massachusetts, Police Sergeant Patty Fisher, "It happens every year at this time. Residents in the area of Sunshine Dairy Farm may notice loud noises coming from the dairy cows at all hours of the day and night. We've been informed that the cows are not in distress and that the noises are a normal part of farming practices."[291] First, I would ask, informed by whom? Certainly not by an animal behaviorist. While separating baby calves from their mothers may be a routine practice for a dairy farm, it is unnatural. In fact, according to a 2015 article, removing a calf within hours of birth interrupts the maternal bonding process and creates less sociable adult cows.[292]

Another study that investigated how dairy calves respond emotionally to the practice determined this kind of traumatic incident can, like any PTSD, result in the cow having a "negative cognitive bias."[293]

Within hours of being born, the calves are separated by gender. The female calves are put in isolation hutches, away from their mothers. The males are a waste product in the dairy industry, so they are either killed or sent to a veal farm.

The "humane" and "best-practice method" to kill the male calves is to prod them down an increasingly narrow hallway or path so there's no escape by the time they get a sense of what is in store for them. Once they arrive, they're shot in the head with a gun that sends a metal rod into their brain. This rod doesn't kill them but merely stuns them to make the remainder of the process "humane." While they are still stunned, a person throws them up on a table where another person slits their throat. Then, while still alive, someone tosses them up on a rack to dangle by one foot where they finish bleeding out. Again, this is the "humane" method.

Thankfully, I've never been to a slaughterhouse to see this process in person, but I have seen videos of it online, and it's horrifying.[294] This isn't a one-off video; you can go down a rabbit hole because there are so many slaughter videos like that one. It's revolting to see stunned and dying newborn calves, the most innocent of creatures, tossed around with complete indifference, like suitcases at an airport. Be warned: some things you can never unsee.

Once a young heifer is old enough to become impregnated and begin lactating, she is connected to milking machines and lives out a life of exploitation. Normally a cow's average life span can be twenty years or more, but the life expectancy of a dairy cow is four to five years. At that point, after repeated pregnancies, dairy cows can no longer produce enough milk to be profitable, so they are taken to the slaughterhouse. Sadly, by then, some are so weak that they can no longer stand, let alone walk to their deaths.[295]

We've talked about the dairy industry's inherent cruelty, but what about its environmental impact? A mature dairy cow weighing 1,400 pounds generates fourteen gallons of feces and urine every day. The total excrement for a one-hundred-cow dairy housed in free-stall total confinement is "approximately 9 tons and 9+ cubic yards of manure daily."[296]

Remember that all of this extreme cruelty and waste is necessary for the Meatrix to provide us with a glass of milk or a piece of cheese. However, many plant-based milk and cheese products are readily available without any of the cruelty or the biohazardous waste.

Sadly, the egg industry is just as cruel, perverse, and over the top.

Within the Meatrix a chicken has two uses: meat and eggs. "Broilers" are used for meat and "layers" are used for eggs. Through selective breeding practices, broilers' bodies rapidly grow, while layers develop more slowly. Broilers are both male and female, while layers are only female.

Our world has seen steady growth in animals slaughtered for meat over the past fifty years. For instance, in 1961, 172.96 million cattle were slaughtered for consumption, and in 2018 the number increased to 302.15 million. For pigs, the increase was from 376.37 million in 1961 to 1.48 billion in 2018. These are significant increases but pale in comparison to chickens. In 1961, the Meatrix slaughtered 6.58 billion meat chickens, compared to a whopping 68.79 billion in 2018—an increase of over 945 percent.[297] Keep in mind that these figures are only broiler chickens and do not include poultry exploited in egg production.

NUMBER OF ANIMALS SLAUGHTERED FOR MEAT, WORLD, 1961 TO 2018 (BILLIONS)

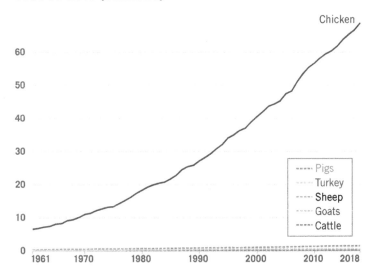

Source: Hannah Ritchie and Max Roser, "Meat and Dairy Production," *Our World in Data* (2017), ourworldindata.org/meat-production

With so many broiler chickens being slaughtered for meat each year, the living conditions and well-being of these creatures are significant concerns of animal rights and animal welfare groups. Broiler chickens reach slaughter age at four to seven weeks of age. These chicks grow so fast and have such large breasts that they can suffer from a myriad of welfare conditions such as skeletal malformation, skin and eye lesions, and congestive heart disease.

Broiler breeders the Meatrix uses to produce more broiler chickens have much longer life spans than their brothers and sisters who are bred for meat and never reach maturity. Poultry farmers allow breeders to mature so they can impregnate them and create more broilers. Because they're allowed to mature, broiler breeder chickens suffer from a different set of ailments. As a result of selective breeding, broiler breeder chickens are seriously food-motivated—basically, they want to eat all the time. If left to their own devices, 20 percent would eat so much they would be unable to move or would die of severe heart conditions. The Meatrix can't profit from dead chickens, so it only feeds them a quarter of what they'd eat on their own with unrestricted food access. As a result, these chickens are always hungry, which leads to harmful pecking disorders.

Feather pecking is a significant problem with broilers and layers when kept in non-cage environments. When one chicken pecks at another's feathers, sometimes removing the feather, the pecked chicken can bleed and have skin irritation, which leads some ravenous chickens to cannibalize them. Ironically, consumers demanding cage-free meat and eggs ask farmers to raise poultry in open pens, increasing feather pecking, which creates greater suffering.

To prevent uncontrollable pecking, the Meatrix has resorted to beak trimming. Calling it trimming makes it sound benign, like clipping one's fingernail, but it's simply another Meatrix euphemism. Beak trimming is a partial amputation of the chicken's beak, which contains multiple nerve endings. The beaks are cut either by a blade or infrared beam without anesthesia. This routine and cruel act helps prevent feather pecking but comes with its own set of problems. Like homing pigeons, a chicken's beak contains structures (iron-containing sensory dendrites) that allow it to navigate using the earth's magnetic field. Research has shown that beak-trimming reduces a chicken's capacity to navigate using magnetoreception, preventing it from moving efficiently between foraging and roosting sites in cage-free environments.[298] Beak trimming also prevents chickens from successfully grooming themselves, leading to a higher rate of ectoparasites than non-trimmed chickens.[299]

Beak trimming is not the only mutilation routinely performed on broiler breeder chickens to help reduce pecking injuries. They may also undergo spur removal, toe trimming, and comb dubbing. These procedures are usually done with a hot wire and, like beak trimming, without pain relief or anesthesia.

With chickens, the artificial insemination process is just as artificial and invasive as it is with bulls and cows. For example, a worker restrains and massages a male broiler breeder's abdomen to arouse it. Then with the other hand, the chicken is squeezed so that semen is ejected and collected. To impregnate a female broiler breeder, a worker inserts a syringe through the oviduct and into her vagina and injects the semen.

In this book, it's impossible to discuss all the ethical problems that arise with the farming of animals. Additional information is available online should you choose to do further research on your own, which I encourage you to do. Knowledge is power. A great place to start is the documentary *Dominion*. A warning, though: it's heartbreaking to watch.[300]

This book aims to inform and inspire you to make better choices, not do a complete exposé of the Meatrix. But, I feel it's crucial you know some of the many ways the Meatrix exploits animals. I imagine most meat eaters haven't given this much thought. I know I didn't think about it when I was part of the Meatrix. I hope this information helps you see it's the Meatrix that is extreme, not plant-based eaters!

The Meatrix pacifies us with bucolic images of farm life. Our entire lives, we see pictures of happy cows, pigs, and chickens on food trucks delivering dead animal parts to our groceries and on the animal products we buy, even though, as mentioned earlier, the public is against animal suffering. Hopefully, educating yourself about the processes the Meatrix uses to whitewash its actions and brainwash you will motivate you to make better choices with your dollars because whether you like it or not, you are complicit in the suffering of animals anytime you purchase an animal product.

FASCINATING FACTS

- In 2019 the US accounted for 46 percent of the world's bull semen market, worth $208 million.
- 68.79 billion chickens were killed for meat in 2018, an increase of over 945 percent since 1961 levels.
- A mature dairy cow weighing 1,400 pounds generates fourteen gallons of feces and urine every day.

CHAPTER 9

Tradition, Taste, and Convenience?

As if it were yesterday, I can vividly recall a fishing trip my father took my brother and me on one morning. I had been looking forward to our boys-only outing all week.

We left very early that Saturday morning and drove about thirty minutes to a place called Guist Creek. If you visit today, you'll see a marina, but fifty years ago, I recall it being nothing more than a lake, nature, and the three of us. There was fog that morning in the low-lying areas that the rising sun gradually burned away.

My excitement for the bonding experience I was about to have with my father and brother quickly turned into trepidation when I learned we were using live bait. I tried to bait my hook, but for some reason, the worm my dad assured me was unable to feel pain wasn't cooperating, making it nearly impossible. My father patiently demonstrated how to do it, reassuring me that this wasn't harming the worm in the slightest, as they didn't have nerve endings like humans did. Unfortunately, his words rang hollow; they didn't jibe with what I could see right before me with my own eyes. Based on its gyrations, the worm appeared to be very much in pain. Why could my dad not see it? Was he oblivious to the harm we were creating?

Eventually, I cast into the shimmering lake. Surely, we had now advanced to the fun part of the experience. But actually, there was little to do but shoo the flies and wait. So far, I couldn't see what all the fuss was about when it came to fishing. Obviously, I was missing it, or it hadn't happened yet.

Suddenly I felt a tug on the line, and my father coached me through reeling the fish in. I was excited because I had successfully caught my first fish and was about to experience firsthand the part of fishing that my dad and many others found so rewarding.

This experience would make hurting the worm worth it. When I was at the point of lifting the fish out of the water to see what I'd caught, my father took over holding the pole and said, "Son, take hold of the fish." I was eager to be getting to the fun part of this expedition and reached out to grab the flopping fish.

When I finally grabbed ahold of it, it was cold, slimy, squirmy, and scaly. The movement of the scales against the palm of my hand startled me like an electric shock. I immediately let go of the fish and let out a squeal, and I quickly retracted my hand, shook it and wiped it on my pants leg.

If anything, my father was persistent; even after all of that, he was determined to salvage the experience and share the joys of fishing. He showed me that the fish was too small to keep, and we'd need to release it back into the lake, but first, we needed to remove the hook from the fish's mouth. By this time, I was horrified but gathered all the strength I could muster to put on a brave face as he demonstrated how to de-hook the bulge-eyed, gasping fish and release it back into the water.

I have no other memories of that day. I don't recall how much longer we stayed or whether my father or brother had any success. All I knew was I was done with fishing, and I have never fished since. As we packed up and headed back home, I was very perplexed. For the life of me, I couldn't see why people thought fishing was fun. What was I missing?

I honestly have no idea how the experience was for my brother, but looking back, I imagine it was probably a disappointing day for my father. We never talked about that fishing expedition, and my father never invited my brother and me on another.

While I may have disappointed my father that day, I certainly was confused by him and his behavior. On what basis could he know that the worm and fish weren't in pain? What enjoyment could he, or anyone, get from killing innocent worms and fish? Maybe my problem was I asked too many questions and didn't blindly accept my dad's explanations.

I open this chapter with that story because passing down hunting and fishing traditions within families helps normalize and romanticize the harm we do to animals. Was my father initiating my brother and me into a tradition that was significant for him? Was he hoping it would be the beginning of a lifetime of family fishing trips? If so, he must have been disappointed that it never became a tradition we would share.

TRADITION

> The less there is to justify a traditional
> custom, the harder it is to get rid of it.
> —Mark Twain

We've lived with gender and racial discrimination for years, even though most people say it's wrong. Likewise, just because certain practices have been allowed, rationalized, and normalized, such as the traditions of hunting, fishing, ranching, trapping, and eating meat doesn't mean they should continue. It might be different if we needed animal products to survive, but we don't.

Sadly, the indoctrination of many ranchers and trappers into the Meatrix begins at a very early age. As a result, many simply follow traditions passed down to them from earlier generations. But things can and do change.

One huge example is the Ringling Brothers and Barnum & Bailey Circus. Due to pressure from consumers, the American Society for the Prevention of Cruelty to Animals, and other animal rights organizations, this ninety-eight-year-old circus ended the performance of elephants on May 1, 2016 (nearly two years ahead of their planned retirement in 2018) and then shuttered operations in 2017.[301]

In October 2021, the COO of Feld Entertainment, which bought the circus from the Ringling family in 1971, said the circus, billed as the Greatest Show on Earth, would relaunch in 2023 without animal performers.[302] If the circus can evolve from its longstanding tradition of using wild animals to entertain us, other industries in the Meatrix can do the same. But as Mark Twain said, it's often difficult to rid ourselves of traditions we can no longer justify.

A case in point is trappers in Montana who, on November 8, 2016, helped defeat a bill that would have placed restrictions on animal trapping on public lands throughout the state. Among the arguments *in support* of the bill was that "trapping is indiscriminate, commercial, cruel and dangerous. It weaponizes public lands, makes them unsafe for pets and non-target wildlife, and encourages disproportionate, for-profit use of a public resource." Among the arguments *against* the bill were that trapping is "part of our heritage that we treasure" and it "dates back to the time of Lewis and Clark. It is a cherished family tradition like hunting, fishing, and camping. Let's keep it that way."[303]

Is trapping really a cherished family tradition like camping?

On a national level, an anti-trapping bill, the Refuge from Cruel Trapping Act of 2021, which, as of March 2022, has forty-seven co-sponsors, was referred to the House Committee on Natural Resources, where it remains. The bill "prohibits, with specified exceptions, the use or possession of body-gripping traps in the National Wildlife Refuge System."[304]

The same day that bill was introduced, the Animal Welfare Institute published a press release stating that 79 percent of Americans support prohibiting trapping on national wildlife refuges, with 88 percent agreeing that habitat preservation should be the system's top priority. Currently, nearly half of the 566 wildlife refuges allow "the tradition" of trapping.[305]

Another time-honored tradition occurs each year on cattle ranches in the US when months-old baby cows are separated from their mothers for the first time and then hogtied to be vaccinated, hot-iron branded, dehorned, and castrated. All of these practices are done routinely without anesthesia or any kind of pain relief. A 2011 *Wall Street Journal* article reported that "the USDA wants every cow to have a unique numerical I.D. tag instead to improve tracking and beef safety." However, ranchers viewed this as a way of vilifying and demoting branding.[306]

Ranchers say these practices are not only about cattle identification but that branding, dehorning, and castrating cows is also a "family get-together time"—a ranching tradition. One rancher defended branding because it was "about community, friends and family...helping neighbors and enjoying a good time together." Ranchers say cattle tagging and other innovations in these ranching practices symbolize the loss of a social tradition—one they say they enjoy and have the right to continue.[307]

The traditions of trapping and ranching, like farming, are often passed down from one generation to the next. So it's understandable that ranchers and trappers might be resistant to changing their ways. But as mentioned earlier, even the circus has evolved to not rely on wild animals to entertain us. If an institution such as the circus can change with the times, I believe trappers and hunters can too.

Families, too, cling to traditions. For instance, traditions such as meat being a central focus of most holiday celebrations are things most Americans enjoy. However, these occasions often create challenges for those living outside the Meatrix. It can be easy to succumb to the societal pressure to "fit in" at such gatherings and fall back into the Meatrix. The good news is it's never been easier to participate in these holiday meals and remain plant-based. You can find a veganized recipe of nearly every traditional holiday dish online. I have taken many plant-based foods to family gatherings over the years.

Another tradition in the US features outdoor grilling and backyard barbeques on holidays during the warmer months. These, too, can be easily managed with some pre-planning because an expanding variety of meat-free, plant-based grilling options now exist. My favorite is Beyond Meat's Beyond Burger, and the company also makes plant-based sausages that cook up nicely on a grill. Of course, corn on the cob, veggie kabobs, and foil-wrapped potatoes all cook just fine on a grill too. Grilled fruits such as mango and pineapple are delicious and provide an unexpected twist to vegan ice cream or a salad.

When I was growing up, our cookouts concluded with roasting marshmallows for dessert. Then, as an adult, I discovered the deliciousness of s'mores, which consist of one or two campfire-toasted marshmallows with a layer of chocolate sandwiched between two graham crackers. While each of these products typically contains animal ingredients, the good news is plant-based versions of marshmallows, graham crackers, and chocolate are available today, so you can still enjoy s'mores once you've escaped the Meatrix!

TASTE

When we have overwhelming evidence that consuming animal products contributes to poorer health and shorter lifespans, why do we continue to eat within the Meatrix? As mentioned earlier, doing so was never a conscious choice for most of us; it was something learned at a very early age and continued without much thought. The unconscious eating of animal products in this context makes perfect sense. We were all born into the Meatrix, so animal products' consumption is second nature to us.

While it's true most people never consciously chose to eat animal products, I think a majority continue to do so because they simply enjoy the taste. We've long known that the appeal of certain foods, like ice cream and chocolate, stems from a powerful combination of sugar and fat.[308] But what drives our obsession with the flavor of meat? Many Americans crave the taste of bacon, a slice of cheesy pepperoni pizza, or a juicy cheeseburger, even though it's common knowledge they will pay the price with weight gain and poor health. So what makes food in the Meatrix so hard for so many to resist? The answer, along with the fat, is the flavor called umami and the Maillard reaction.

Umami

There are four commonly known tastes: sweet, sour, salt, and bitter. But there is a fifth, less identifiable taste that has a potent effect on us: umami (Japanese for "savory deliciousness"). The Umami Information Center describes umami as a pleasant savory taste. These tastes are imparted by an amino acid, glutamate (found in certain cheeses like Parmesan, soy sauce, and many types of meat), and ribonucleotides, including inosinate (found in meats and fish), and guanylate (found in sundried tomatoes, boiled potatoes, edible seaweed, and mushrooms).[309] In recent years, chefs have coined the term "u-bomb," which refers to a combination of foods with two or more of these ingredients that greatly magnify the umami profile of a dish, making it more irresistible. A few examples of "u-bomb" dishes are a pizza with sausage and cheese; mashed potatoes with butter and mushroom gravy; and lasagna made with a combination of sausage, mushrooms, sundried tomato sauce, and cheese.

Even so, when asked, most people would not likely identify a pleasant-tasting dish as having "umami." In fact, it wasn't until the year 2000 that we discovered special umami receptors in our taste buds.[310] Before that, umami was considered only as a flavor enhancer, not a distinct taste in its own right. Some scientists believe nature wired us to be attracted to umami as an indicator of the presence of protein in foods, helping us make choices for this vital nutrient.

Fat

A 2019 report challenged the long-held belief that meat set the stage for human evolution. Its lead researcher, Jessica Thompson, proposed it was more likely the acquired taste for calorie-dense marrow scavenged from the large bones of animal carcasses that were responsible. In fact, the meat of wild animals was too lean to sustain our ancient ancestors: "It actually takes more work to metabolize lean protein than you get back."[311]

One of the most appealing meat components isn't the lean muscle tissue meat but its fat. Paul Breslin, a nutritional sciences professor at Rutgers University, says, "When you think you're craving meat, most likely what you're really craving is fat."[312] This is because when cooked, fats stored in the meat oxidize, wafting into the air and making many a carnivore weak in the knees. By contrast, taste tests of meat created in the lab using stem cells from a cow's shoulder but containing no fat were unappealing. So next time you're jonesing for a juicy cheeseburger or steak, keep in mind that what you're really craving is the fat, not the meat.

Maillard Reaction

The Maillard reaction is named after Louis-Camille Maillard, an early twentieth-century French physician and chemist. In *Meat Products Handbook*, Gerhard Feiner describes the Maillard reaction as "an extremely complex process" that is "the reaction between reducing sugars and proteins by the impact of heat."[313] Like umami, the Maillard reaction isn't unique to meat and is partly responsible for our attraction to non-meat foods such as crusty baked loaves of bread, cookies, and roasted coffee.

However, the Maillard reaction's presence in cooked animal flesh is a powerful driver in our obsession with it. For instance, while our taste preferences are, in many ways, cultural and something that is passed down to us from earlier generations, there's also a powerful biological component to what smells appealing to us that may be the driving force behind our attraction to the Maillard reaction.

Scientists say our response to the Maillard reaction may be evolutionary and an indicator to our brains that the meat before us has been cooked and therefore is safe to eat. This complex chemical reaction from heat kills dangerous bacteria, changes beef's color from pink to brown, and creates the alluring aroma of grilled meats and sautéed foods. In fact, up to 95 percent of what we consider the taste of meat is actually aroma, according to Barb Stuckey, author of *Taste: Surprising Stories and Science About Why Food Tastes Good*.[314]

But there's a price to be paid for this love affair with the Maillard reaction. The chemical term for the Maillard reaction is glycation, which produces advanced glycation end products (AGEs), some of which are unstable and therefore reactive, affecting the tissues and cells in our bodies. We have mechanisms in our bodies for eliminating AGEs, but we also have AGE receptors (RAGEs) that attract and store AGEs (contributing to bone cell metabolism). The storing of these AGEs can create persistent cellular inflammation. So it's somewhat paradoxical that the same cooking that kills deadly bacteria, rendering meat edible to us, and enhances the smell and taste of some foods also makes those same foods more damaging to our bodies.[315]

Low levels of AGEs in the body are considered normal, but high levels lead to oxidative stress and inflammation, increasing the risk for many diseases. Diet is the number one contributor to the accumulation of AGEs in the body. While you can't avoid them except through a raw diet, you can reduce your intake by avoiding foods (most of which also happen to be animal-based) with the highest amounts of AGEs. These foods are meat (especially red meat), certain cheeses, fried eggs, butter, cream cheese, margarine, mayonnaise, oils, and nuts. Fried foods and highly processed products also contain high levels of AGEs.[316]

High fructose corn syrup (HFCS)—an ingredient found in many processed foods like commercially made candies, chips, cookies, and crackers—increases AGEs in the body as well. Healthier brands without HFCS are available, but you can also choose whole-foods snacks, such as fruit. Studies also show that regular exercise and a diet high in antioxidants found in plant-based foods further reduce AGEs and inflammation in the body.[317]

The great news is by simply escaping the Meatrix you'll significantly reduce the amounts of harmful AGEs in your diet. According to healthline.com, you can lower them even more by cooking slower, at lower temperatures, and using moist cooking methods such as stewing, boiling, poaching, and steaming.

ADVANCED GLYCATION ENDPOINTS FOR VARIOUS ANIMAL AND PLANT FOODS

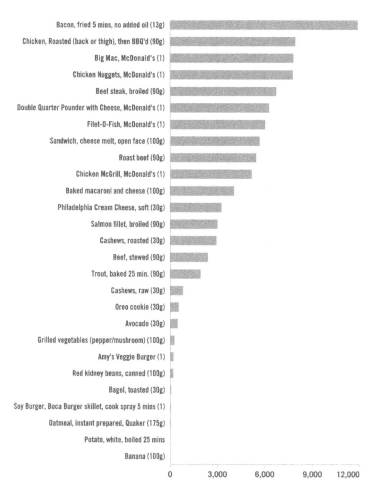

Source: Jaime Uribarri et. al., "Advanced glycation end products in foods and a practical guide to their reduction in diet," *Journal of the Academy of Nutrition and Dietetics* 110, no. 6 (June 1, 2010): 911-16, www.ncbi.nlm.nih.gov/pmc/articles/PMC3704564/

CONVENIENCE

The "convenience" argument was probably valid in the past. However, because of the rapid growth of plant-based foods coming to market, the convenience argument today is flimsy at best. Certainly, there are far too many areas in our country without nearby grocery stores, and I support efforts to rectify that. But for those fortunate to have a local supermarket, plant-based products are increasingly available. Plant-based restaurants are easier to find than ever, but even if your town doesn't have one, you can still go out with your friends and order a salad and baked potato, even at the most hardcore steakhouse. As my mother was fond of reminding me, "Where there's a will, there's a way," and that applies here too.

If you're committed to eating outside the Meatrix, you can make it work. The good news is it requires less and less effort because, as mentioned earlier, the number of people choosing a plant-based lifestyle has increased exponentially in recent years. More grocery stores, fast food outlets, and restaurants are catering to them, so the convenience argument is quickly becoming less and less viable.

We all need to eat. We eat multiple times a day. Because of that, we rarely think about what we're putting in our mouths. Instead, we seek out what tastes good or what's quick and handy. We eat unhealthy, animal-based foods on holidays to fit in with others or relive special childhood memories. But, as we've seen in this chapter, making food choices based solely on tradition, taste, and convenience is not only unhealthy; it's downright harmful—to you, the planet, and animals. Thankfully, plant-based eating is becoming more accessible; it's easier than ever before to have delicious, cruelty-free food any time, even when grabbing a quick meal on the go.

FASCINATING FACTS

- 79 percent of Americans support prohibiting trapping on national wildlife refuges.
- 88 percent of Americans agree that habitat preservation should be the national wildlife refuge's top priority.
- 95 percent of what we consider the taste of meat is actually aroma.
- The meat of wild animals was too lean to sustain our ancestors.

CHAPTER 10

Cognitive Dissonance

I vividly recall a conversation I had when a group of family members and I were carpooling to a wedding in Cincinnati, Ohio. I had requested a plant-based meal for the wedding reception, and my brother-in-law, who was driving, asked me why I was plant-based. The car suddenly became utterly silent. It was as though everyone inhaled at the same time, sucking all the oxygen out of the vehicle as they braced for my answer. The discomfort in the car was palpable. I was grateful he asked, and thankfully, I had just read *Diet for a New America* by John Robbins, and I happily shared some of the harm that animal agriculture does to the earth, animals, and human health. That was it. There was no further inquiry; the discussion just dropped. It was as if I had been speaking in a foreign language no one could understand. No one in my family had ever asked me that question, and no one has ever bothered to talk to me about it since. As we'll discuss in this chapter, the lack of curiosity and avoidance is a dissonance-reducing strategy. Ignorance is bliss.

Cognitive dissonance is the discomfort we feel when our actions are not in alignment with our beliefs. When the two are in alignment, our choices become simple. As Bugs said to Morpheus in *The Matrix: Resurrections* when offering him the choice between the red and blue pill, "The choice is an illusion. You already know what you need to do."

A classic example of cognitive dissonance is smokers who know that smoking is bad for them. This discord between belief and action is known as cognitive dissonance. So what happens with the smokers who believe it's unhealthy to smoke, or a person who knows eating meat is inhumane, bad for the environment, and harmful to their health? Well, some take action and quit, while many others continue to smoke or eat meat even though they know all the reasons they shouldn't.

As mentioned earlier, a 2012 study revealed that 94 percent of Americans are against doing unnecessary harm to animals.[318] Yet most Americans engage in unnecessary harm three times a day. The "dissonance" part of cognitive dissonance is like the operating system (OS) that runs constantly in the background of your computer. Unfortunately, if you're like most people, you give your computer's OS very little attention until you get a virus or have some trouble with it.

In the same way, we pay little attention to the many ways our actions aren't in alignment with our beliefs. It's not until something or someone makes us think about the conflict between our beliefs (we know we should be eating healthier) and our actions (stopping at McDonald's on the way home from work) that we can no longer ignore the discrepancy between our beliefs and actions. This sudden cognition makes us aware of the dissonance, and we become uncomfortable and sometimes even defensive, and we attempt to justify our actions or lash out in anger. I've seen it happen time and time again in conversations between plant-based activists and non-vegans.

In real life, most people don't think about the animals they eat at all, and the Meatrix wants it that way. It seems to me that those who *do* consider farm animals must view them as nothing more than biological machines, indifferent to their lives.

One proponent of this line of thinking that heavily influenced western culture was seventeenth-century French philosopher Rene Descartes. Descartes, widely considered the father of modern philosophy, denied that animals had intelligence or reason. Instead, he viewed animals as soulless and therefore incapable of feeling pain or anxiety. To him, animals were "automata," biological machine-like creatures that functioned similarly to more simplistic machines like watches and cuckoo clocks. While not everyone accepted his views of animals as automata, generally Western cultures adopted them, leading to the mistreatment of animals with impunity, sanctioned by laws and societal norms up to the mid-nineteenth century. It wasn't until Charles Darwin's theory of evolution that our convenient and self-serving ideas of animals being no more than machine-like creatures began to change.

A 2021 *New York Times* article by columnist Farah Manjoo explored whether animals (in this case, his cat) have consciousness. In the article, Manjoo says if we believe our cats and dogs have consciousness, then it becomes incredibly difficult to deny that other creatures, like farmed animals, have consciousness, too. He writes,

> Yet if all these creatures experience consciousness analogous to ours then one has to conclude that our species is engaged in a great moral catastrophe—because in food production facilities all over the world, we routinely treat non-human animals as Descartes saw them, as machines without feeling or experience. This view lets us inflict any torture necessary for productive efficiency.[319]

While the concept of animals being automata is comforting to anyone living in the Meatrix, it bears no resemblance to the truth. In fact, neuro-scientific discoveries reveal animals, including farmed animals, have much more in common with us than previously thought.

Recent advancements in science have created the need to reassess our long-held beliefs about non-human animal consciousness. In 2012, a group of prominent international scientists gathered for the Francis Crick Memorial Conference on Consciousness in Human and non-Human Animals at the University of Cambridge to do just that. The panel of cognitive neuroscientists, neuropharmacologists, neurophysiologists, neuroanatomists, and computational neuroscientists published "The Cambridge Declaration on Consciousness," declaring that "the weight of evidence indicates that humans are not unique in possessing the neurological substrates that generate consciousness. Non-human animals, including all mammals and birds, and many other creatures, including octopuses, also possess these neurological substrates."[320]

In 2021, the London School of Economics and Political Science announced that Britain's Animal Welfare (Sentience) Bill protecting animal welfare had been extended to include all decapods, crustaceans, and cephalopod mollusks. This independent, government-led commission investigated over three hundred scientific studies. As a result, it concluded that cephalopods (including octopuses, squid, and cuttlefish) and decapods (including crabs, lobsters, and crayfish) should be "regarded as sentient, and should therefore be included within the scope of animal welfare law."[321]

Additional research reinforces the growing scientific evidence that you're not eating "some things" but "someones." A 2019 article by neuroscientist and animal behavior and intelligence expert Lori Marino explains how the latest research shows that farm animals are not things and that many, even under the incredible stress of factory farming, exhibit some of the same characteristics we recognize within ourselves and in the pets we love and let into our homes. Like us and our pets, farm animals have distinct personalities and complex lives we're beginning to recognize.[322]

A 2011 study by the University of Bristol's School of Veterinary Sciences established that domestic hens exhibit empathy, physiological, and behavioral responses when their chicks are mildly distressed. They explained that they "used chickens as a model species because, under commercial conditions, chickens will regularly encounter other chickens showing signs of pain or distress due to routine husbandry practices or because of the high levels of conditions such as bone fractures or leg disorders."[323]

Of course, the Meatrix isn't concerned about its stressful husbandry practices, high incidences of leg disorders, or bone fractures. It views these as the typical cost of raising animals as food commodities and simply allows these and other ailments to go untreated. However, always looking out for its bottom line, the Meatrix is greatly concerned about other chickens encountering these sick and crippled birds because studies indicate that stressed animals taste bad. The Meatrix has so thoroughly documented this that Marta Zaraska, author of the book *Meathooked* (which I highly recommend), devotes an entire not-for-the-faint-of-heart chapter to this one subject.

The latest scientific research into farm animals' complex social, emotional, and intellectual lives is a real eye-opener to the average person. It refutes the idea of farm animals as elementary, biological machine-like creatures with no concern for their own lives or the lives of their family and friends. Instead, Lori Marino's article concludes that we resist the evidence because we don't want to "deprive our palate and shatter our sense of ourselves."[324]

In 2014, I had the good fortune of traveling around southern Spain during Semana Santa (Holy Week) with my nephew. We spent a couple of days in the lovely city of Granada and went for a walk one night after dinner. As we made our way through one of Granada's barrios, I noticed some vegan graffiti, which was reassuring because we had found Spain to be primarily a country of meat eaters. The graffiti said, "*Los animales no son nuestros esclavos. Hazte vegano. Stop especismo. Son y pueden ser tus amigos.*" The translation is, "Animals are not our slaves. Become vegan. Stop speciesism. They are and can be your friends." My nephew and I chatted a bit about this graffiti and how encouraging it was. During our conversation, he shared a story about his grandfather, who was a farmer. He said his grandfather became so attached to the animals he raised on his farm that it was difficult for him to send them to the slaughterhouse. These animals became more like pets than farm animals.

The discomfort this avid farmer experienced shows another example of cognitive dissonance, as well as an example of farm animals being someone instead of something. Based on what my nephew said, his grandfather was conflicted about killing the innocent animals he raised, nurtured, and grew to recognize as individuals similar to pets.

I consider his grandfather as another victim of the Meatrix. He believed he should behave a certain way as a farmer but, based on my nephew's account, was troubled by doing so.

Another story illustrating cognitive dissonance or the lack of it occurred in China. Sensibilities in China about dogs are changing as its growing middle class can now afford to have them as pets. In 2011, a Beijing truck was transporting dogs to slaughter for meat, which is not uncommon throughout Asia. A microblogger spotted the truck and posted it on social media, asking animal lovers to help him rescue the dogs. About two hundred people showed up. The bravery of these animal lovers and their activism touched me, but one of the things I remember most about the article was a quote from the truck driver, who said, "I transported dogs as

VEGAN GRAFFITI IN GRANADA, SPAIN

(I would) pigs, cows, and sheep. The country does not ban the consumption of dog meat." While the driver saw no difference between dogs and other farm animals, a growing number of people in China do.[325] I believe the uncomfortable feeling many people have about eating dogs should also be extended to other animals because the line between a dog and a farmed animal is an arbitrary one created by speciesism and the Meatrix.

Yet another example of cognitive dissonance happened in Florida in 2015: a man was sentenced to one year in jail for intentionally running over nine ducklings with a lawnmower. It was an egregious act—for sure—but if there's going to be a public outcry (even jail time) about intentionally killing innocent animals, then it seems to me there should be a public outcry about intentionally killing all innocent animals. The convicted man's behavior seemed cruel and irrational because it's not normal for someone to purposely go out of their way like that to kill innocent ducklings.[326] But essentially, that's what happens at every meal if you're a part of the Meatrix. People who are outraged by animal abuse but continue eating within the Meatrix remind me of myself when I was a meat eater. I was naïve, unaware, and complicit. I said I was against needless animal suffering but didn't realize the tremendous suffering that all food animals endure. The Meatrix wants us to believe that farmed animals are somehow different. That they were born to be our food, and this somehow rationalizes the suffering these animals experience. But suffering is suffering. Injustice is injustice. Exploitation is exploitation. Period.

We can have the highest intentions, thoughts, and words about what we eat and how we live our lives, but we remain in a hypocritical and disempowered state until our actions support those intentions, thoughts, and words. I think so many resist discussing their diet because they're trying to avoid the uneasy feelings, especially the guilt, from knowing better but not acting better, that cognitive dissonance creates.

Hank Rothgerber of the Department of Psychology at Bellarmine University in Louisville, Kentucky, found that simply being in the proximity of a plant-based eater agitates cognitive dissonance within omnivores. In fact, the dissonance is so palpable that meat eaters engage in various dissonance-reducing strategies, including avoidance, dissociation, perceived behavioral change, denial of animal pain, denial of the animal mind, pro-meat justifications, and more.[327] Dr. Rothgerber's studies show that these tactics are nothing more than rationalizations meant to assuage the uncomfortable dissonance omnivores face within themselves when in proximity to a person who lives without killing animals and eats only plants.

Plant-based activists often encounter people within the Meatrix who exhibit many dissonance-reducing strategies. For example, I've watched online videos where people claim they need meat for many reasons science disagrees with, such as combating anemia and building strong bones and muscle, as well as men who think eating meat will raise their testosterone levels and enhance their libido. None of these statements are based in science, yet they are lodged firmly in some of the minds of meat eaters as a way to justify their eating habits.

One of the more interesting examples I've seen is a video in which a man shares his belief that a cow is honored when we eat it and then go out and do good things in the world. Irrespective of our diet, I believe we should all strive to do good for the world. Doing kind and loving acts is honorable and aspirational. But we can do these things without ever harming an animal.

To me, ignoring the pain and suffering animals endure by saying that cows are honored to be our food when we do good deeds in the world because of the energy they provide us is another dissonance-reducing strategy. It's perhaps not a complete denial of animal pain, but it certainly minimizes it and rationalizes that the ends somehow justify the means.

Dismissing the harsh reality animals face in slaughterhouses reminds me of how in *The Matrix*, Agent Smith attempts to convince Neo that the alternate reality provided to humans by the AI somehow more than adequately compensates humans for their bondage to the Matrix. I think most movie viewers were appalled by this line of thinking, just as I believe cows would be appalled upon hearing a person state that the cows would consider it an honor to be killed and eaten by humans.

As a person who follows a plant-based lifestyle, I often hear rationalizations from non-vegans about why they continue to eat meat, and the Meatrix does all it can to support these dissonance-reducing justifications. One such example is a well-known phrase touted by many ranchers: "Cows have a great life and one bad day."

Does a cow raised for beef really have a great life and only one bad day? Cattle often are branded with hot irons, creating third-degree burns. Additionally, they routinely have their ears tagged, their horns cut or burned off, and their scrotums clamped so tightly their testicles atrophy (if they are not outright castrated)—and all of this is routinely done without pain relief or anesthesia.[328]

While on the range, few cattle receive adequate veterinary care, and some die from neglect or injury. In fact, nearly 3.9 million cattle died from predator and nonpredator causes in 2015, a figure that has remained fairly consistent since 2000.[329] Additionally, cattle can freeze to death in states like Nebraska, South Dakota, and Wyoming, while others may die from heat exposure in more southerly states like Kansas and Texas.

After a year of harsh weather conditions, cows are rounded up and packed so tightly they are usually unable to move around. They're shipped to an auction house and then sent perhaps hundreds of miles away to live by the thousands in cramped, muddy, and feces-filled feedlots. Here, they are fed an unnatural diet primarily consisting of corn, which fattens them up quickly but can also lead to painful bloating and acute acidosis. Furthermore, feedlot cattle often suffer from respiratory diseases due to breathing the toxic fumes of the feedlot air. In fact, respiratory problems account for the highest nonpredatory deaths in both calves (26.9 percent) and cattle (23.9 percent).[330] In addition to a high corn diet, these cows are often given drugs to fatten them up more quickly. Finally, once these cattle reach slaughter weight, they are forcibly packed into trucks once again and transported to the slaughterhouse, where they experience their "one bad day."

Focusing on the "one bad day," animal rights activist and author Carol J. Adams writes:

> As one person points out: Calling it "one bad day" is a horrendous insult in itself. This isn't the same as hitting your toe on the coffee table or spilling coffee on yourself in the car. This is the equivalent of being dragged out of your house and shot in the head with a 20-gauge shotgun. This isn't

something that will be better with a good night's sleep or an aspirin.[331]

People often ask vegans what will happen to all the animals if we stop eating them. To me, this line of questioning is another dissonance-reducing strategy. Some people within the Meatrix conveniently believe we're somehow doing animals a favor by killing and eating them, as if we're thinning their herd for them. But nothing could be further from the truth. The animals in the Meatrix only exist in the numbers they do because of consumer demand. I believe as the world becomes more plant-based, our demand for meat will decrease. As a result, the Meatrix will gradually create fewer animals until the profit margins are too low to be profitable. Like most things in life, it's about supply and demand, which is about making money.

FASCINATING FACTS

- In 2012 the Cambridge Declaration on Consciousness stated that many animals possess the neurological substrates that generate consciousness—including all mammals, birds, and octopuses.
- 300 scientific studies concluded that cephalopods (including octopuses, squid, and cuttlefish) and deca-pods (including crabs, lobsters, and crayfish) should be regarded as sentient.
- A Florida man was sentenced to one year in jail for intentionally killing nine ducklings.
- Four different studies found that simply being in the proximity of a plant-based eater agitates cognitive dissonance within omnivores.

CHAPTER 11

The USDA Dietary Guidelines

Every five years, the Dietary Guidelines Advisory Committee (DGAC), part of the USDA, is responsible for reviewing the latest scientific evidence and making recommendations on how Americans should eat. However, the 2020 DGAC contained Meatrix industry insiders: "More than half the committee members come with either clear strings to industry-funded research or questionable memberships in industry-funded advocacy groups and foundations."[332]

It may surprise you that the committee's recommendations seem to change little regardless of who is in the White House. In fact, when First Lady Michelle Obama was planting an organic garden in the White House lawn and encouraging people to eat their fruits and vegetables, the DGAC continued to be pro-Meatrix.

But for the first time, the 2015–2020 guidelines included recommendations for sustainable eating, saying that "a diet higher in plant-based foods, such as vegetables, fruits, whole grains, legumes, nuts, and seeds, and lower in calories and animal-based foods is more health-promoting and is associated with less environmental impact than is the current US diet."[333] This recommendation was a victory for everyone who seeks to be healthier and who cares about the future of our planet.

Nevertheless, there was fierce blowback by the Meatrix.

Whether one chooses to believe what scientists tell us about human activity and climate change, on this one thing most would agree the science is definitive: the Meatrix is bad for the environment. However, not surprisingly, congressional leaders doing the bidding of the Meatrix stripped the sustainable eating recommendation from the guidelines. Despite strong opposition from nutritional groups, the following was tacked on to, of all things, an appropriations bill, saying the DGAC's guidelines are explicit and "limited in scope to nutritional and dietary information."[334] The intense pushback to the dietary guidelines' comments about the environmental benefits of plant-based eating indicates that clearly, the Meatrix does not want Americans to consider the environmental impact of the food on their plates.

With so many industry insiders serving on the Dietary Guidelines Advisory Committee, it's understandable how this conflict of interest makes it difficult to update guidelines in a way that is founded strictly in science. In the past, when the committee wished to alert people of the health dangers of consuming animal products, their recommendations were often vague, pro-industry, and full of scientific jargon that most people didn't understand. For instance, instead of advising people to eat less meat, they recommended Americans lower their saturated fat intake. This carefully crafted wording favors the Meatrix because many Americans don't know what saturated fats are or that all animal products contain them. Another tactic is advising we "eat lean meats," which still encourages the consumption of meat. Encouraging Americans to reduce "saturated fatty acids by replacing them with monounsaturated and polyunsaturated fatty acids" is an example of jargon. In plain speak, this recommendation means eat less animal-based foods and replace them with more plant-based foods. But so far, meat industry members on the DGAC have not allowed the guidelines to have such clear-cut wording, and it's not likely to change anytime soon.

In reality, the DGAC guidelines matter little in how most Americans eat. Still, they're significant because they dictate federal programs covering school lunches, prison meals, the Supplemental Nutrition Assistance Program (SNAP), and more. According to the USDA Economic Research Service, in 2019, the National School Lunch Program provided 29.6 million free or low-cost lunches to students daily for $14.2 billion annually.[335] The Meatrix gets a large piece of this pie as long as the USDA's dietary guidelines include animal products. The DGAC guidelines are supposed to be rooted in the latest scientific research, and the science is definitive: animal products are harmful to your health. If industry insiders didn't make up more than half of the DGAC, the science would likely win out, and the Meatrix would lose big.

As already mentioned, few Americans follow the DGAC's guidelines, with most people exceeding the committee's recommendations for animal protein and saturated fat every day. For example, according to David Robinson Simon, author of the book *Meatonomics*, teenagers consume 78 percent more saturated fat and 48 percent more cholesterol—both of which are primarily found in animal foods—than the DGAC recommends for their age group.[336] In addition, the Dietary Guidelines 2015–2020 reported that both male and female Americans who are eight years of age and older are eating fewer vegetables and fruits than recommended.[337]

But more alarming and of more significant concern is the data on animal protein intake. The 2015-2016 *What We Eat in America* report—the only nationally representative survey of total food and beverage consumption in the US—shows that all males aged twelve and older exceed the daily recommended protein intake by an average of 167 percent while men ages nineteen to thirty-nine exceed the amount even more by nearly 189 percent per day![338] According to the 2020 DGAC report, "meat and poultry are the predominant contributors of protein foods among adults," with an animal to plant-based protein ratio of 86/14 respectively.[339] Heart disease is the leading cause of death for men in the United States, killing 357,761 men in 2019—that's about 1 in every 4 male deaths.[340] As I mentioned in Chapter 3, based on a study by Dr. Esselstyn of the Cleveland Clinic, the only diet proven to reverse heart disease is plant-based.[341] So based on these two pieces of information alone, shouldn't a plant-based diet be the de facto diet of US men ages twelve to seventy?

Based on the above, it seems clear Americans pay little attention to the official dietary guidelines and exceed the DGAC's recommendations in some key areas, such as with animal protein and saturated fats. Therefore, it was surprising that the 2015–2020 report indicated that the average American consumes *less* dairy than the DGAC recommends. This conclusion is astonishing when you consider that cheese consumption in the US has already increased from 11.4 pounds per person annually in 1970 to a whopping 40.2 pounds per person in 2019. This is an increase of more than 254 percent![342]

If eating more than forty pounds of cheese per person per year isn't enough to meet Americans' daily dairy requirements, what *are* the requirements, and what are they based on?

The current guidelines recommend Americans consume three eight-ounce servings of dairy daily in the form of milk, cheese, yogurt, or other dairy products. However, Americans aren't eating nearly the amount of dairy the guidelines recommend (and thank goodness because forty pounds of cheese

U.S. CONSUMPTION OF CHEESE CONTINUES TO RISE (POUNDS PER PERSON/YEAR)

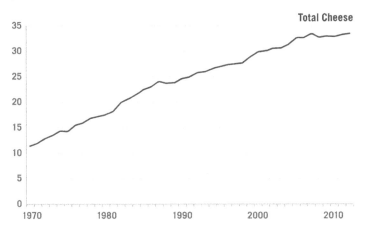

Source: Jeanine Bentley, "Trends in U.S. Per Capita Consumption of Dairy Products, 1970-2012," *Amber Waves* (June 2, 2014), ers.usda.gov/amber-waves/2014/june/trends-in-us-per-capita-consumption-of-dairy-products-1970-2012

per year alone is a lot of dairy).[343] In fact, according to a 2017 bulletin by the USDA, "On average, Americans consumed 1.5 cup-equivalents of milk and milk products (49 percent of the recommended three cup-equivalents) per day in 2014."[344] So how can it be that Americans are eating far more dairy today than ever (thirty pounds more cheese per year than in 1970) but still consuming less than half the recommendation made by the DGAC? It's because the DGAC is a pro-Meatrix, industry-insider-laden committee.

According to a *New England Journal of Medicine* article from February 2020, "a central rationale for high lifelong milk consumption has been to meet calcium requirements for bone health." But the article continued by saying, "Paradoxically, countries with the highest intakes of milk and calcium tend to have the highest rates of hip fractures."[345] And yet, according to the same author in a different publication, the DGAC's recommendation for three eight-ounce servings of dairy daily is "perhaps the most prevailing advice given to the American public about diet in the last half-century."[346] So imagine my amazement when I read that the DGAC's guidelines stem from a series of studies assessing the calcium intake and output of 155 adults for two to three weeks![347]

I'd say these short studies don't meet the criteria for making broad recommendations about an entire nation's dietary consumption of any product. For example, clinical trials for drugs typically require three phases to determine if they can be approved for use. According to the NIH, a phase I trial "tests an experimental treatment on a small group of often healthy people (twenty to eighty) to judge its safety and side effects and to find the correct drug dosage." Phase II consists of a larger group of people (one hundred to three hundred) emphasizing effectiveness. The NIH reports, "This phase aims to obtain preliminary data on whether the drug works in people who have a certain disease or condition. These trials also continue to study safety, including short-term side effects. This phase can last several years." Phase III trials include up to three thousand test subjects and gather "more information about safety and effectiveness, studying different populations and different dosages, using the drug in combination with other drugs." If the FDA agrees that the phase III trial results are positive, it approves the experimental drug or device.[348]

I wondered if more extensive bone density and mineralization studies using larger populations and more extended periods had been conducted. The answer is yes. The six-year study referenced in the *NEJM* article reported that "among nearly 10,000 men and women representative of the US population, calcium intake was unrelated to bone mineral density at the hip."

Back to the DGAC's guidelines on dairy intake, how is it possible to determine if lifelong, daily consumption of a particular food group will lead to increased bone density and fewer hip fractures in a matter of weeks? I would say it's not, and as you just read, science agrees. In condemning the DGAC's results, the *NEJM* article stated, "trials lasting one year or less can be misleading, and the two-to-three-week balance studies used to establish calcium requirements have limited relevance to fracture risk."[349]

While saying we all need three eight-ounce servings of dairy daily is great for the dairy industry's profitability, it's bad news for consumers' health. It offers no protection against hip fractures, and as I mentioned previously, scientific studies show dairy consumption is linked to increased rates of heart disease, prostate cancer, ovarian cancer, diabetes, obesity, and other conditions. (See Chapter 3.)

FASCINATING FACTS

- 78 percent of American teens exceed the dietary guidelines for saturated fat.
- Men in the US ages 12–70 exceed the dietary guidelines for protein consumption.
- Heart disease is responsible for 24.2 percent of American male deaths each year.
- The average person in the US consumes 40.4 pounds of cheese a year.
- Recommendations for dairy consumption in the US were determined by a two- to three-week study of 155 people.
- 10,000 men and women studied for six years determined that calcium intake was unrelated to bone mineral density at the hip.

- 100,000 people followed for more than twenty years indicates that milk consumption is associated with higher mortality and hip fracture.

The Meat Myth, Marketing, and Low Prices

Saying meat has protein is like saying cocaine makes you happy. While both may be true, there are many other things to consider when consuming either product.

THE MEAT MYTH

In the past, many people ate meat because they thought they needed animal protein. Prior to 1838 when scientists first identified the macronutrient "protein," many people incorrectly believed that we needed to eat animals because "flesh makes flesh." We know better now. And while there's a mountain of evidence from decades of scientific research that indicates eating flesh is detrimental to our health, the idea that we need to consume meat is still very prevalent in our society.

Before 1838, most people ate meat not only because they believed their bodies needed it but also because they enjoyed the taste and considered it a status symbol, much like we do today.

For instance, a 2018 article by the University of Technology Sydney reported that "it was a desire for status that drove preference for meat, rather than other variables such as hunger or perceived nutritional benefits."

Further, it stated that "there is a symbolic association between eating meat and strength, power and masculinity. It is traditionally a high-status food, brought out for guests or as the centrepiece of festive occasions, so we wanted to better understand this link to status."[350]

Few food items represent affluence and status more than meat. The United States saw a steep rise in affluence and consumerism after World War II, when people began purchasing meat at increased rates. Therefore, it should surprise no one that, according to the American Chicken Council, Americans in 2020 consumed fifty-eight pounds more meat and poultry annually than they did in 1960.[351]

Globally we're eating a lot more meat today than we used to, and as mentioned in the previous chapter, throughout their entire adult lives, American men eat more meat protein than the DGAC's daily recommendation. But what are the origins of those recommendations for protein?

The foundation of our protein guidelines today is from observations made in 1877 by Carl von Voit (a German physiologist) and his protégé, Wilbur Olin Atwater.[352] According to *Meatonomics*, Voit's recommendation for eating meat was based purely on the observation of how much men of various professions ate, regardless of whether they should consume that amount or not.[353] Voit "recommended 118 g/day for an adult man of average weight, even though he had determined that 52 g/day was enough (later research showed even lower levels were enough)."[354]

By contrast, in 1888, Danish nutritionist Mikkel Hindhede recommended dietary changes based on his conclusion that Danish people consumed too much meat. Hindhede's research established that twenty-two grams to twenty-five grams of protein was sufficient to not only maintain the body's nitrogen balance but even store one gram of nitrogen in the body.[355]

Hindhede's research was put to the test in 1917 when Denmark ran low on foods due to a World War I trade blockade. The government created a committee composed of Hindhede and four farmers who were charged with finding a way to support the dietary needs of the Danish people during the war. The committee decided to reduce pig and cattle herds and use the harvested food, originally intended to feed livestock, to feed humans instead (a recommendation previously made in this book). Hindhede compared the mortality rate of men twenty-five to sixty-five years of age from October 1917 to October 1918 to the seventeen previous years. He discovered that the mortality rate was 34 percent lower. Hindhede was careful not to conclude causation but believed mortality rate reduction supported his conviction that Danes ate too much meat.[356]

Based on our Chapter 3 discussion of Dr. Esselstyn's findings of reversing heart disease in as little as three weeks by eating only plants, it's certainly plausible that a year of eating more plants and much less meat could have been responsible for the Dane's significantly reduced mortality rate.

But, as it turns out, people love hearing good news about their bad habits, so there's little wonder that a population eager to eat more and more meat, regardless of scientific studies, gravitated toward Voit's recommendations of elevated levels of meat over Hindhede's. Thus, the protein myth was born.

However, the meat myth pertains not only to the amount of protein needed for optimal health but also to its source (animal vs. plant).

When people find out I'm plant-based, many ask me where I get my protein. All foods have various nutrients, and all proteins contain variable amounts of amino acids that the body uses to build and maintain muscle. There are twenty amino acids, but nine must be obtained through diet and therefore are considered essential. The Meatrix has many people believing it has a monopoly on these nine essential amino acids, but it doesn't.

Fortunately, an article published in July 2021 reported that while it's true that amino acids do vary in individual foods, the amino acids obtained through diet vary little between those who eat meat and those who eat only plants. The study concluded, "Our data show relatively little difference in protein quality between plant-based and omnivorous dietary patterns and that reduced total protein intake in plant-based dietary patterns may be a contributor to the benefits of plant-based diets." Thus, this latest research not only bursts the meat-myth bubble regarding the superiority of animal protein but concludes that one of the benefits of plant-based eating is the overall healthier protein consumption.[357]

Unlike fats and carbohydrates, our body cannot store the protein we eat. Kristi Wempen, a Mayo Clinic Health System registered dietary nutritionist, says that "once needs are met, any extra is used for energy or stored as fat."[358] Therefore, eating more protein than our body needs at any given time is wasteful. This is an important consideration for US males ages twelve and up, who, as previously mentioned, eat more protein than recommended. For a while, I was one of them!

As an experiment in my mid-fifties, I decided to give a low-carb, high-protein/high-fat diet fad a try. As a result, I religiously weighed and measured my plant-based food before every meal and used an app on my phone to keep track. I also slowed my running down considerably to force my body to burn fat as fuel.

I followed this regimen for about a year, and I lost most of my body fat.

To maintain this elimination diet, I drank three protein shakes a day—and easily ingested 180–200 grams of plant-based protein daily in smoothies alone! I also cut out as many carbs as possible and ate a lot of healthy plant-based fats. Even though I was pumping iron four days a week and running on my off days, this extreme way of eating gave me a healthy-*looking* body but no increase in muscle mass. More importantly, I began to wonder what such an extreme elimination diet was doing to my overall health.

According to a 2009 article from the *Journal of the Academy of Nutrition and Dietetics,* eating an excess of "30 g protein in a single meal does not further enhance the stimulation of muscle protein synthesis in young and elderly."[359] Yet three times a day, I was drinking more than twice that in each shake!

So what does thirty grams of animal protein look like on a plate? There are approximately seven grams of protein per ounce of cooked meat, so an eight-to-nine-ounce beef steak is fifty-six to sixty-three grams of protein, about twice the amount medical science says our bodies can make use of in one meal. The average boneless/skinless chicken breast weighs six to eight ounces, which is forty-two to fifty-six grams of protein. If consuming more than thirty grams of protein at a time has been scientifically proven to be unbeneficial for muscle growth, why do it?

During my time on this diet, I saw my primary care physician, and he ran some bloodwork. While my previous two lipid panels from 2012 and 2014 were stellar and entirely in the optimal range in every category, the 2017 panel was significantly different in some key areas. For instance, my triglycerides had increased to 146 mg/dl—still below the 150 mg/dl target, but just barely. Also, my total cholesterol-to-HDL level, which is a measurement of all cholesterol in the body compared to HDL ("good" cholesterol), had increased by more than 26 percent to 2.4, well within the recommended <5.0 but still a dramatic increase. Nonetheless, these changes make me wonder what would have happened to my cholesterol and overall health if I'd continued following such a restrictive diet.

Moving away from protein shakes and weighing my food, I moved into what I feel is a more natural way of eating and living. I've found that it's not difficult for me to get my daily allowance of protein with a plant-based lifestyle and that occasionally calculating my dietary reference intake (DRI) of protein is helpful and simple. The DRI was introduced in 1997 as a system of nutrition recommendations from the National Academy of Medicine of the National Academies. The daily reference intake of protein is 0.36 g of protein per pound (0.8 g per kg) of body weight. As mentioned earlier, the average for a sedentary man is 56 g and for a sedentary woman is 46 g.[360]

As a plant-based person, I no longer obsess about getting enough protein (or any other nutrients), but I am mindful of how much I'm eating. If you're just starting on your plant-based journey (Yay, you!), or are already plant-based and wonder if you're getting enough protein, I recommend using an app to track your macronutrients—fats, carbohydrates, and protein. The only one I've used (because I find it easy) is MyFitnessPal.

I decided to track my meals the other day to see how much protein I was eating. On this particular day, I ate oatmeal with pecans, sunflower seeds, and hulled hemp hearts for breakfast (17.3 g of protein). For lunch, I ate lentils with extra virgin olive oil, fire-roasted tomatoes, capers, and raw pumpkin seeds (21 g of protein). For dinner, I ate curried butternut squash with coconut milk, fire-roasted tomatoes, garbanzo beans, green peas, and cashews (31.8 g of protein). Finally, I ate raw almonds as an afternoon snack (12 g of protein). So my total protein intake for the day was 82.1 g, exceeding the daily reference intake formula (mentioned above) for an average male by 26.1 g. Since I weigh twenty-five pounds less than the average white US male in my age group, my margin above the DRI was even higher. If you're active and want to bump up your plant-based protein intake, you can easily add things such as tofu, tempeh, edamame, or seitan to a recipe. If, for some reason, you feel this still doesn't give you enough protein, you can include a plant-based protein bar or shake sometime during your day. So you see how easy it is with a bit of planning to meet or exceed your daily reference intake of protein eating only plants.

As I mentioned earlier, I no longer drink protein shakes or weigh, measure, or count any nutrients. But I am conscious about eating a wide variety of plant-based foods to obtain the amino acids I need to build muscle and enjoy the many other health benefits I get from eating a rainbow of colors. Plus, it tastes great!

In December 2021, I saw my primary care doctor for some bloodwork, and after a few years of giving up protein shakes and simply eating plant-based foods, my lipid profile was again well within the "optimal" range in every category. My total cholesterol-to-HDL ratio had dropped back down from 2.4 to 2, and my triglyceride level had dropped by 98 mg/dl to the lowest point of any previous test.

I no longer sacrifice my health for the appearance of fitness. Instead, my research indicates that we're setting ourselves up for improved health, greater happiness, and increased longevity simply by escaping the Meatrix and eating whole, plant-based foods while engaging in moderate exercise.

The 2009 article mentioned before isn't the only study to dispel the notion of plant-based proteins being inferior or incomplete. A December 2016 study reported, "Protein from a variety of plant foods, eaten during the course of a day, supplies enough of all indispensable (essential) amino acids when caloric requirements are met."[361] So as long as you're eating enough calories, your protein needs can be met with only plants.

A *Nutrition Reviews* article from February 2019 also discussed the type and quality of plant-based and animal-based proteins, stating, "A typical day of eating a common variety of foods includes adequate amounts of both essential and nonessential amino acids, almost regardless of the presence or absence of animal foods."[362] A different article from 2021 corroborated this, stating that "protein source did not affect changes in absolute lean mass or muscle strength."[363]

The 2019 article also included the following graph of the amino acids found in various animal and plant-based foods. While there are subtle differences, the amounts of amino acids, particularly the essential ones that we must obtain through diet, are, in many ways, similar regardless of the source.[364]

MARKETING

It's unclear how much animal protein we would eat without the Meatrix's relentless marketing, as well as the artificially low prices of animal-based foods. So first, we'll take a deeper look at how the Meatrix markets its products to us.

The Meatrix works hard advertising its products to us, hoping we'll increase our consumption. There are many more fruits

Proportions of amino acids in selected foods across food groups. Amino acids are grouped as essential or nonessential, in descending order of prevalence within food groups. Amount of protein per 100 kcal is presented. (Source: Nutrition Database System for Research, University of Minnesota; http://www.ncc.umn.edu/ndsr-database-page/). *Abbreviations*: Ala, alanine; Arg, arginine; Asp/n, aspartate and asparagine; Cys, cysteine; Glu/n, glutamate and glutamine; Gly, glycine; His, histidine; Iso, isoleucine; Leu, leucine; Lys, lysine; Met, methionine; Phe, phenylalanine; Pro, proline; Ser, serine; Thr, threonine; Trp, tryptophan; Tyr, tyrosine; Val, valine.

PROPORTIONS OF AMINO ACIDS IN SELECTED FOODS ACROSS FOOD GROUPS

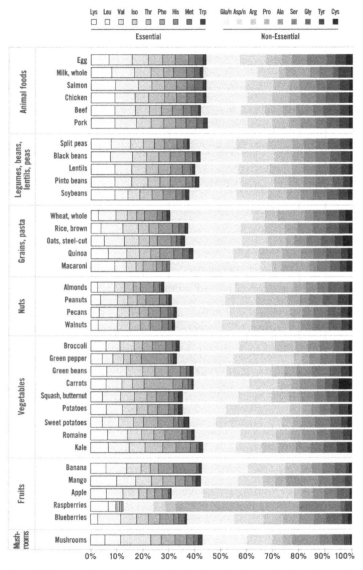

Source: Christopher D. Gardner, et. al. "Maximizing the intersection of human health and the health of the environment with regard to the amount and type of protein produced and consumed in the United States." *Nutrition Reviews* 77, no. 4 (April 2019): 197-215, doi.org/10.1093/nutrit/nuy073

and vegetables in the grocery store than meats, so why are we not inundated with fruit and vegetable marketing campaigns? The answer is a particular type of promotion the Meatrix is highly effective at using to its advantage: checkoffs.

Initially, various industries offered checkoffs, allowing producers to opt in to this unique type of promotion by paying a fee. Checkoffs do not single out any individual producer or brand; still, checkoff leaders hope the additional advertising generated by the checkoff program will lead to greater visibility and increased demand so that everyone in the industry will benefit—employing the concept that "a rising tide raises all boats." These programs were originally voluntary but eventually became mandatory—a tax, so to speak, that some farmers cannot escape.[365]

How much money do checkoffs generate? In the US, each beef producer pays $1 for every cow sold, and every pork producer pays $0.40 for every $100 of value. In Canada, the checkoff is $1 per animal; in Australia, it's $5. According to *MeatHooked*, the beef checkoff in the US collected $1.2 billion between 1987 and 2013.[366] That's quite a stockpile of funds dedicated to the sole purpose of generating increased demand for the Meatrix's products. Is there any wonder why we consume more meat than we need?

Of course, there are checkoffs for non-Meatrix products as well. In fact, according to the USDA's Agricultural Marketing Service, half of the twenty-two USDA research and promotion programs that checkoff monies fund are for fruits and vegetables, including pecans, Hass avocados, blueberries, mushrooms, mangos, peanuts, potatoes, watermelon, sorghum, and soybeans. However, according to *Meatonomics*, "the dairy industry spends more on advertising in one week than the blueberry, mango, watermelon, and mushroom industries spend together in a year."[367]

Although most people have never heard of checkoffs, nearly everyone is familiar with their catchy and effective ad campaign slogans:

- "Got Milk?"
- "Beef. It's What's for Dinner."
- "Milk. It Does a Body Good."
- "Pork. The Other White Meat."
- "The Incredible Edible Egg."

These slogans are catchy and effective. According to industry reports, the beef checkoff increases sales by nearly twelve dollars for every dollar spent, and the pork checkoff collects more than thirty-eight dollars of value to the producer for every dollar spent on research and more than twelve and fourteen dollars for indirect and direct promotions respectively. The lamb checkoff holds its own, generating anywhere from fourteen to thirty-eight dollars for every dollar spent.[368]

Clearly, checkoffs are highly profitable and effective in getting us to eat more and more meat, but are they popular with farmers, or even constitutional?

Several challenges to checkoff programs began early in the twenty-first century. In August of 2000, thirty thousand pork farmers voted 53 percent to 47 percent in favor of terminating the mandatory pork checkoff "tax." However, then-Secretary of Agriculture Ann Veneman never published a final ruling, meaning she effectively refused to complete the program's termination. Hog farmers of the Campaign for Family Farms claimed that industry lobbyists, such as those from the National Pork Producers Council (NPPC), worked as advisors to the secretary of agriculture. As Missouri hog farmer Rhonda Perry noted, "The problem is, Veneman is listening more to NPPC lobbyists than to hog farmers in the countryside."[369]

Hog farmers were not the only ones trying to end mandatory checkoff programs. In 2001 the US Supreme Court ruled that the checkoff program for mushrooms violated the Constitution's First Amendment. Since then, several lower courts have ruled mandatory checkoff programs were unconstitutional. However, in May of 2005, a very interesting thing happened within the checkoff programs when the US Supreme Court agreed with beef checkoff defenders, ruling that the beef program did not violate free speech because it was government speech (a position not considered in the mushroom case).[370]

But how can mandatory checkoffs be government speech if the revenue is paid by farmers to the government? Has the government taken over these citizens' free speech?

It turns out that the Department of Agriculture reviews all the slogans and catchphrases of the Meatrix's checkoff programs. This led author David Robinson Simon to write, "It may say the National Pork Board, but the background sounds you're hearing are the imposing bass tones of the US government...a lack of government involvement would likely lead to the decline—or maybe the end—of checkoffs."[371]

Why is the government so involved with the Meatrix's checkoffs? Again, the answer is money. As mentioned earlier, according to the American Meat Institute, the US meat and poultry industry generated an estimated $1.22 trillion of economic activity in 2019[372], or roughly 6 percent of the entire gross domestic product (GDP). As mentioned earlier, the beef checkoff alone collected $1.2 billion between 1987 and 2013, but not all this money was spent promoting beef.[373]

A March 2017 article reported on the incestuous relationship between the pseudo-governmental boards that control checkoff funds and the federal government's oversight of them. The article brought to light the past abuses of checkoff funds and a bipartisan effort, Senate bill S.741 (2017–2018), to bring greater transparency to these campaigns.[374]

One example mentioned in the article is how the United Egg Board attacked the plant-based company Hampton Creek (now named Just). Hampton Creek successfully launched a plant-based mayonnaise called Just Mayo. In 2016, the United Egg Board spent checkoff funds "to halt the growth of Hampton Creek. The campaign included paying consultants to try and get Whole Foods to stop carrying Hampton Creek products and paying bloggers to discredit the company."[375]

Other abuses of checkoff funds include the lobbying group the NPPC (the same that prevented the termination of the pork checkoff mentioned above), which in 1997 used checkoff funds to spy on animal rights activist groups. In addition, the National Cattleman's Beef Association misused $216,944 in beef checkoff funds the USDA forced them to reimburse.[376]

Passing Senate bill S.741 could bring greater transparency to how checkoff funds are spent, but unfortunately, it hasn't seen the light of day. The bill was introduced in the Senate on March 28, 2017. It was read twice and sent to the Committee on Agriculture, Nutrition, and Forestry, where it remains today.

Checkoff programs have also taken a play right out of Big Tobacco's playbook by donating large sums of money to universities that then conduct industry-favorable research. A perfect example of this appeared in an October 2021 article that reported that meat eaters "on average experience lower levels of depression and anxiety." However, this article buried some crucial information in the final paragraph: "the data was insufficient to investigate causal relationships" between depression, anxiety, and meat eating.[377] Fortunately, the study itself includes the following disclosure: "This study was funded in part via an unrestricted research grant from the Beef Checkoff, through the National Cattlemen's Beef Association."[378]

Through years of brainwashing, catchy advertising, and sponsored research, the Meatrix convinced me (and likely most of us) that meat and its products were necessary. People stuck in the Meatrix no longer need to be persuaded, just occasionally prodded to remember that meat is the king of the plate—even if it's the pauper of your health and the health of the planet. The Meatrix, like Big Tobacco, simply needs to sow a seed of doubt in consumers' minds by creating messages contrary to nearly all nutritional science. The Meatrix hopes these messages will hide what's real to consumers and prevent them from making better-informed and healthier choices in the grocery aisle. It reminds me of something Bugs tells Neo in *The Matrix Resurrections*: "If you don't know what's real, you can't resist." So to prevent consumers from resisting it, the Meatrix attempts to hide what's real.

LOW PRICES

In addition to the Meatrix's industry-funded research and advertising campaigns, I'd like to show you how the artificially low prices the Meatrix creates keep us coming back for more.

Another argument I hear against forgoing animal-based foods is the expense of adopting a whole-foods, plant-based lifestyle. However, it's easy to eat a low-budget plant-based meal. Although the processed vegan foods at the supermarket tend to cost more than their animal-based counterparts, beans, rice, and potatoes are pretty inexpensive.

It's essential to keep in mind that prices of animal products are low because the government subsidizes many of the industries producing these products.

For example, the USDA gave $9 billion to beef producers in 2020 (in both direct and indirect payments). By comparison, lentil growers received $15.5 million the same year, meaning for every $1 allocated for lentils, cattle ranchers received more than $470 in benefits. The ratios are similar for other Meatrix products. For instance, for every $1 spent on mushrooms, $160 went to pork producers, and for every $1 directed to oats, $80 went to dairy.[379]

These government handouts to the Meatrix tip the scales of the "free market" in its favor. As a result, many vegan foods tend to cost more upfront, though many plant-based eaters argue they save money in the long run due to improved health and fewer trips to the doctor. For example, I've found that I visit the doctor's office less frequently now than when I was a vegetarian (that means I ate dairy and eggs, but I didn't eat meat).

Wealthy industrialized countries like the US provide massive subsidies to farmers, leading to market distortion. The lower prices of subsidized foods influence our shopping habits, resulting in more people buying less expensive foods such as beef, poultry, milk, and other subsidized Meatrix products. It's a boon for the Meatrix. The subsidies are funded by taxes, so citizens are ultimately footing the bill while the industries making the products reap all the profits. Consequently, foods that would typically be costly (such as animal products) become cheap, while foods with smaller or no subsidies remain expensive by comparison.

According to the *San Francisco Chronicle* in 2001, "If water used by the meat industry were not subsidized by taxpayers, common hamburger meat would cost $35 a pound. You need 25 gallons of water to produce a pound of wheat—2,500 gallons to generate a pound of meat."[380] Curious what impact water prices in my hometown might have on the cost of a pound of beef, I discovered the Kentucky American Water Company's price in 2019 for 1,000 gallons of water for farm customers was $17.87.[381] Multiply $17.87 for 1,000 gallons by 2.5 (for the 2,500 hundred gallons of water needed for one pound of beef), and the water cost alone would be $44.68 for every pound of beef. When one considers everything, not only the water, required to bring that one pound of hamburger to you, the actual price per pound of beef is mind-blowingly high.

In *The Value of Nothing: How to Reshape Market Society and Redefine Democracy*, author and activist Raj Patel discusses research done in India to determine the actual costs of a four-dollar hamburger. Researchers found that depending upon whether rainforest land was used somewhere in the process—which would lead to loss of carbon and decreased biodiversity—a burger could cost as much as $200.[382]

This cost distortion leads to people—particularly those with lower incomes who cannot afford more costly alternatives—making food choices that are not in their best interests. As I mentioned earlier, the government should reconfigure its subsidies to incentivize growing more plants for humans, rather than animal consumption. Increasing the subsidies for plant foods for humans would drive down the costs of eating healthily and incentivize farmers to move away from producing animal products.

Speaking of farmers, contrary to what some may believe, vegans love farmers! We want them to be happy and prosperous, just not at the expense of our planet's health. Indeed, farmers in the Meatrix will need assistance transitioning to a plant-based world, just as any evolving industry—like solar and wind power generation to combat climate change—has had help. Of course, as more people free themselves from the Meatrix, there will be less demand for animal products, so farmers will slowly transition to growing more plants and fewer animals. Hopefully, at the same time—or sooner—government subsidies will shift away from animal agriculture to more healthy, sustainable, plants-for-humans agriculture.

Necessity is the mother of invention. One can look at the current food industry to see how the Meatrix might respond to changes in the market. In the 1990s, drinking habits began changing as the media increasingly (and incorrectly) blamed the obesity crisis on soda consumption. As a result, by the late 1990s, soda consumption was on the decline. How did Coca-Cola respond? In the 2000s, it bought water companies, sports drink companies, tea companies, and more. Tyson Chicken, for another example, in 2016 bought stock in plant-based meat company Beyond Meat.[383] Why would Tyson Chicken invest in a plant-based competitor? The answer is diversification.

Diversification is how the farmers of the Meatrix will survive too. The government could introduce changes in the subsidy program that would incentivize farmers to retool toward a plant-based world.

One needs to look no further than nut-milk company Elmhurst to see how farmers could pivot to plant-based operations. According to its website, Elmhurst was founded in 1925 to deliver hand-bottled milk to Queens and Brooklyn, New York, residents. In 2016, Henry Schwartz, son of Elmhurst's founder, closed operations. He reopened in 2017 in Buffalo, New York, as a plant-based milk company. Again, the Elmhurst website says the decision to retool Elmhurst wasn't an easy one but a forward-thinking one. Henry saw the growing trend away from cow's milk to plant-based alternatives and decided the switch was not only good for the environment but good for his bottom line. As a result, according to its website, Elmhurst is the fastest-growing plant-based milk company on the market today.[384]

According to the Centers for Medicare and Medicaid Services, healthcare costs in the US grew 9.7 percent in 2020, reaching $4.1 trillion, or $12,530 per person per year, and accounted for 19.7 percent of gross domestic product (GDP).

Most of the procedures, medications, and tests each year in the US are for chronic diseases that often can be prevented, slowed down, or even reversed by a plant-based lifestyle.[385] Shifting subsidies away from animal products that make us sick to healthier, plant-based foods could help lower healthcare costs by preventing, stabilizing, or reversing many chronic diseases (see Chapter 3).

The subsidies that keep meat inexpensive have created a world the likes of which we've never seen before. Marta Zaraska, in her book, *Meathooked*, chronicled our obsession with meat for the last 2.5 million years. Historically speaking, meat has been a luxury commodity that only the very wealthy and privileged could afford to enjoy on any consistent basis. For most people, meat was scarce and eaten only on rare occasions such as celebrations, weddings, and festivals. Not that long ago we looked back at diseases such as diabetes, heart disease, hypertension, and cancer as diseases of affluence. Today, we have flipped the script so that some of the least expensive foods available are animal-based. As a result, diseases once relegated to the super-rich are now the leading causes of death for the masses.[386]

Thanks to pro-Meatrix government subsidies, it's often less expensive for a family to eat fast food—generally, meat and animal products loaded with saturated fats—than buy fresh plant-based foods and prepare family meals at home. Animal products are no longer luxury items but would be if not heavily subsidized. How many people would choose to pay up to $200 for actual the cost of a hamburger? If prices of animal products weren't artificially low, would our love affair with meat wane? Shifts of pennies in the price of gasoline impact people's driving habits. Changes in prices also affect people in the grocery aisle.

A good illustration of this can be found in *Meathooked*, where author Marta Zaraska reports that when the cost of beef goes up 10 percent, people purchase 7.5 percent less beef, which results in increased sales of pork by 3 percent and chicken by 2.4 percent. Additionally, 35 percent of people surveyed by the National Chicken Council reported buying more veggies when the price of chicken increased.[387] Of course, the Meatrix cannot have people abandoning their products, so it lobbies our government to keep its prices artificially low. However, passing the actual costs of animal products on to the consumer would dramatically shift behavior away from unhealthy animal products toward healthier plant-based choices.

Argentina is a case in point. In 2016, the country was second only to Uruguay in meat consumption. But in January 2020, a report noted that 60 percent of Argentines were considering going plant-based.[388]

Why the sudden shift? Argentina's currency declined by 50 percent in 2018, followed by another 25 percent in 2019. The interest in plant-based eating in Argentina is a perfect example of how prices shape eating habits. The fact that 60 percent of Argentines are considering going plant-based because of economic hardship indicates how closely tied our consumption is to the artificially low prices the Meatrix works so hard to create.

If we're not going to do away with subsidies to the Meatrix, the least we could do is subsidize non-animal products to a similar degree.

FASCINATING FACTS

- Americans in 2020 consumed 58 pounds more meat and poultry annually than in 1960.
- Consuming greater than 30 grams of protein in a single meal does not further enhance the stimulation of muscle protein synthesis in young and elderly.

- The US beef checkoff collected $1.2 billion between 1987–2013.
- The USDA gave $9 billion assistance in direct and indirect payments to beef producers in 2020.
- The estimated true cost of a hamburger when including the loss of carbon and decreased biodiversity is $200.
- In 2020, the cost of healthcare in the US was $4.1 trillion.

CHAPTER 13

Feel Great!

Friends and family may question whether or not plant-based eating is healthy. You can reassure them by letting them know the following:

> It is the position of the Academy of Nutrition and Dietetics that appropriately planned vegetarian, including vegan, diets are healthful, nutritionally adequate, and may provide health benefits in the prevention and treatment of certain diseases. These diets are appropriate for all stages of the life cycle, including pregnancy, lactation, infancy, childhood, adolescence, older adulthood, and for athletes. Plant-based diets are more environmentally sustainable than diets rich in animal products because they use fewer natural resources and are associated with much less environmental damage.[389]

As I've noted throughout this book, simply escaping the Meatrix will improve your health. But just as most meat eaters don't eat for their health, many plant-based eaters don't either. Many accidentally vegan foods, such as Oreo cookies and Fritos, are available, but no one thinks of them as healthy. You can be plant-based and help offset climate change, end the suffering of animals, and alleviate the burden on the environment while still eating junk food! But in order to feel great and take full advantage of all the health benefits eating only plants offers, you need to eat not only for the planet and the animals but also for your health. An easy way to do this is by avoiding processed foods and eating whole fruits, vegetables, legumes, beans, nuts, and seeds. Another way is to eat a rainbow of colored veggies on your plate.

I've got some bad news and some good news about nutritional deficiencies. First, the bad news. Generally speaking, meat eaters eating the standard American diet tend to be deficient in calcium, fiber, folate, iodine, magnesium, vitamin C, and vitamin E.[390]

Now we'll get to the good news. The same study also reported that plant-based eaters tend to have only three nutritional deficiencies: calcium, iodine, and vitamin B12. Since calcium and iodine are also on the list of nutritional deficiencies of omnivores, as a person who has escaped the Meatrix, you really have only one potential nutritional deficiency that's different from those eating animal-based foods.

This is excellent news! Simply escaping the Meatrix offers plant-based eaters fewer nutritional deficiencies than omnivores. Furthermore, it's reassuring to know that the few deficiencies mentioned above can easily be overcome with additional plant-based foods and a bit of planning. This section will discuss a few things plant-based eaters can do to ensure they get even more of the nutrients they need to thrive from the foods they eat.

But first, I want to share with you three easy ways to make sure you're getting enough calcium, iodine, and vitamin B12 while eating plants:

1. Calcium: eat some dark green vegetables every day

2. Iodine: season with iodized sea salt or a pinch of kelp once or twice a week

3. Vitamin B12: eat B12-fortified foods such as nutritional yeast

Boom! It's that easy. You've just addressed the three most common plant-based nutrient deficiencies by eating non-animal-based foods.

But if you would like more information about these three nutrients and more things you can do to feel even greater once you've escaped the Meatrix, read on. Nearly everything below also works if you're eating the standard American diet, so this is *not* a list of extra things you need to do once you've escaped the Meatrix. These are just commonsense things that will make you feel even better regardless of how you eat. The main thing is to keep it simple. It doesn't have to be complicated. By escaping the Meatrix, you've already done the best thing you can do for yourself, the planet, and animals!

Dr. Michael Klaper, who has been practicing plant-based medicine for over forty years, says, "You're not what you eat. You are what you absorb!"[391] With that in mind, let's talk about some simple things you can do to get the most from the food you're eating!

CALCIUM

As already mentioned, calcium is a possible deficiency for both plant-based and meat eaters. In Chapter 11, you learned that one of the most prevalent messages to the American public about diet in the last fifty years is to consume three eight-ounce servings of dairy daily. However, "Calcium from plant sources may even be better for you than calcium from milk or other animal products, since animal proteins leach calcium from your bones." It's essential to make sure you're eating enough fruits, vegetables, and other plant-based foods that contain calcium, including kale, soybeans (edamame, tofu, miso, soy milk, tempeh), bok choy, broccoli, oranges, and seeds (poppy, pumpkin, sesame, chia).[392]

IODINE

Iodine is naturally present at various levels in earth's soils which, in turn, impacts the iodine content of crops. As already mentioned, plant-based and meat-based eaters alike can become nutritionally deficient in iodine. In fact, iodine deficiency severely impacts brain development and has debilitated millions of people around the world. The WHO says, "iodine deficiency is the single most important preventable cause of brain damage."[393] Unfortunately, even a low dietary intake of iodine can lead to hypothyroidism, the symptoms of which are fatigue, weight gain, and water retention.[394]

Thyroxine, T4, is an essential hormone produced by the thyroid that contains four iodine atoms. According to the NIH, the T4 hormone regulates energy metabolism in all cells and is critical for good health.

Daily seasoning of food with iodized salt provides a small and steady amount of sufficient iodine. However, sadly, the average person consumes five to ten times more salt than needed. Many people, on the advice of their physician, are reducing their salt intake. That's because reducing dietary sodium lowers blood pressure and reduces vascular disease. Salt has been called both a public health hazard and the neglected silent killer.[395]

Thankfully, only half a teaspoon of iodized salt daily meets the recommended dose of iodine. But if you're avoiding salt or reducing your consumption of it for health reasons, there's good news. You can get all the iodine you need from plant-based foods! Consuming one tablespoon of sea vegetables such as wakame, arame, and dulse flakes two or three times per week can provide those thriving outside the Meatrix with all the iodine they need. [396]

However, there are a few words of caution. Be careful using kelp because it contains so much iodine you can create the opposite problem—hyperthyroidism. Just a sprinkle of kelp one or two times a week is all that most people will need. It's also best to avoid the sea vegetable hijiki because it absorbs arsenic from the ocean. Finally, don't rely on sea or Himalayan salt alone for iodine. If you're counting on sea salt for your iodine, be sure to get the iodized version of sea salt as iodine evaporates during the sea salt-making process. And while pink Himalayan salt contains trace amounts of iodine, it alone is not an adequate and reliable source of iodine.[397]

VITAMIN B12

While vitamin B12 wasn't on the deficiency list for omnivores, it turns out B12 deficiency is not a plant-based problem. According to the *American Journal of Clinical Nutrition*, "Vitamin B-12 deficiency and depletion are common in wealthier countries, particularly among the elderly, and are most prevalent in poorer populations around the world." The reason B12 deficiency is common among the elderly is because our gastric juices become weaker the older we get, making it more difficult for our bodies to absorb B12 from the foods we eat.[398]

Because a plant-based source of vitamin B12 was unknown until recently, many people in the Meatrix will tell you nature doesn't intend you to be plant-based. However, in November 2019, Parabel, a company that hydroponically grows and processes water lentils into Lentein, discovered their water lentils (more commonly known as duckweed) contain a tremendous amount of natural bioactive forms of B12—100 grams of dried Lentein equals 750 percent of the US recommended daily value! An independent third-party laboratory verified these findings and proved that a plant-based form of vitamin B12 does exist. Perhaps we'll discover even more plants containing B12 in the coming years.[399]

The B12 found in animal products does not come from animals. Instead, vitamin B12 exists in bacteria in the soil, and ruminant animals get B12 from ingesting soil attached to the plants they consume. However, as mentioned in Chapter 4, only 4 percent of the US beef market is grass-fed, so 96 percent of all US beef is not consuming B12 naturally from the soil. Instead, they are being supplemented with cobalt, which synthesizes vitamin B12 in the stomachs of ruminants.[400]

The easiest way to get your recommended daily allowance of B12 is to supplement (crystalline forms of B12 are easily absorbed). Alternatively, some vegan products—nut milks, cereals, and nutritional yeast—like many name-brand, non-vegan breakfast cereals, may be fortified with vitamin B12, making them a reliable source of this vital nutrient. I love the taste of nutritional yeast, and I sprinkle it on lentils, kale, stir-fries, and many other foods, and I know it's good for upping my vitamin B12 intake. However, not all nutritional yeasts contain B vitamins, so if you're relying on it for vitamin B12, make sure yours does.

Some sea vegetables (spirulina, nori, wakame), as well as miso, have a pseudo vitamin B12 that can block the absorption of B12. Vitamin B12 analogs do this by occupying the receptor sites on your tissues that B12 would use, blocking vitamin B12 from being absorbed.[401] Therefore, it's recommended you avoid eating sea vegetables on the days you consume your B12 by either supplement or a fortified food such as nutritional yeast. Doing so will allow the B12 analogs to clear out of your tissues, making the receptor sites available for B12.

CHEW, CHEW, CHEW!

Regardless of what food you put into your mouth, chewing it thoroughly is a great way to get the nutrients from the food by initiating the pre-digestive process. For plant-based eaters, it's important to know that a strong cell wall of cellulose protects plant nutrients. The great news is that because nature intends us to be herbivores, our teeth are perfect grinders for breaking down those tough cell walls. So if we want to get the most benefit from the healthy, plant-based food (really any food) we're eating, we must chew it to a pulp. According to board-certified nutritionist Marilee Nelson, "Taking the time to pre-digest what you eat makes digestion more energy efficient by dramatically increasing nutrient uptake and reducing the workload on the stomach and small intestine."[402] Because of this, I recommend eating consciously and taking a few extra seconds with each bite to thoroughly chew your food.

SOAKING

Nature packs beans, legumes, nuts, and seeds with many nutrients, and there are ways of maximizing their absorbability. According to the NIH, grains, legumes, oilseeds (rapeseed, soy, sunflower), and nuts contain phytic acids, which inhibit the absorption of minerals such as iron, zinc, calcium, magnesium, and manganese. Soaking grains and beans before cooking can be "quite effective for reduction of phytic acid as well as consequent increase in mineral bioavailability."[403]

While phytates will prevent our bodies from absorbing some nutrients, they're not all bad news. For instance, while phytates may interfere with the absorption of some minerals, they also act as potent antioxidants that can slow cell damage and aging. In addition, "Phytates should not significantly impair mineral status when included as part of a diverse and balanced diet, especially if using traditional processing methods such as soaking, germinating, fermenting, and cooking."[404]

SPROUTING

According to *Nutrients*, sprouting (also referred to as germinating) food reduces phytic acid content by 40 percent.[405]

The Cleveland Clinic reported sprouts are jam-packed with beneficial vitamins, minerals, fiber, and are an excellent source of antioxidants. Specifically, they cite nutritionist Mira Ilic: "Broccoli sprouts will be loaded with vitamin A, vitamin C, vitamin K, folic acid, and they are a really good source of the powerful antioxidant sulforaphane."[406]

In addition, a 2020 article noted that eating sprouts:

1. May lower blood glucose levels and help diabetics manage blood sugar levels.
2. May help control insulin levels in the body because sprouted foods contain fewer carbohydrates.
3. Provides insoluble fiber that acts as a prebiotic that feeds the good intestinal bacteria.
4. Can lower cholesterol levels in people with obesity or diabetes.
5. Is often a lower-fat option because food made from sprouts, like tofu and soy milk, has less fat and more protein.[407]

There are four main types of foods used for sprouting: legumes (such as beans and peas), vegetables (such as alfalfa and broccoli), nuts and seeds (such as sunflower and pumpkin), and grains (such as quinoa and wheatgrass).

Sprouts are often eaten raw or lightly cooked and, like all raw foods such as salads, may contain bacteria that can lead to food-borne illnesses. However, there are measures you can take to lower your risk of foodborne illness. You can:

- Avoid commercially grown sprouts and grow your own.
- Keep sprouts chilled below 48 degrees.
- Take extra sanitary measures when sprouting at home.
- Avoid eating slimy or smelly sprouts.
- Always wash your hands when handling sprouts.
- Always rinse sprouts before eating them.

To reduce the risk even further, the Cleveland Clinic recommends cooking the sprouts, adding, "You may lose some vitamins and minerals when you cook sprouts, but you're still getting most of the nutrients they contain, just to a lesser amount."[408]

You can often purchase grains that have already been sprouted at your local grocery; for instance, our local Costco sells organic, gluten-free, sprouted rolled oats and Go Raw's sprouted organic pumpkin seeds with sea salt, and Trader Joe's sells extra firm organic sprouted tofu.

SOUPS AND STEWS

Cooking helps to break down the cell walls of the vegetables, which in turn releases many nutrients into the broth. Just be sure to drink the broth left in the bottom of the bowl!

IRON

Eating vitamin C-rich foods, such as oranges and strawberries, with a meal allows more iron to be absorbed by the body. In fact, "taking 100 mg of vitamin C with a meal increased iron absorption by 67 percent."[409] So feel free to add a squeeze of lemon to your kale or mandarin oranges to your spinach salad. It's important to note that the bioavailability of iron in plant-based foods is reduced due to the phytic acid (mentioned above) these foods contain. So soaking your beans, legumes, nuts, and seeds before eating them adds additional benefits by making iron readily available in those foods.

EPA, DHA, AND ALA

The three main types of omega-3 fatty acids are eicosapentaenoic acid (EPA), docosahexaenoic acid (DHA), and alpha-linolenic acid (ALA). According to an NIH article, omega-3 fatty acids are needed to create brain cells and for other vital functions in the body, such as maintaining heart health and protection from stroke.[410]

Omega-3s are essential (meaning our bodies can't create these nutrients, and they must be obtained from our diets) and exist in various foods. For example, many plant-based foods such as chia seeds, flax, and English walnuts are excellent sources of ALAs that our bodies convert into EPA and into DHA.

However, it's important to note that our bodies are not particularly efficient at converting ALA into EPA and DHA: the NIH reports our body's conversion of ALA into EPA and then to DHA can be less than 15 percent. This means the only practical way for those following a plant-based lifestyle to increase the levels of these fatty acids in the body is to increase consumption of EPA and DHA directly.[411]

While it's true that fish oils contain omega-3s, that doesn't mean fish produce it. Fish ingest EPA- and DHA-rich microalgae and plankton, which are then stored in their tissues. The fact that fish aren't the source of EPA and DHA is excellent news for those wanting to thrive outside the Meatrix and means we can bypass them (and the dangerous mercury, lead, nickel, and cadmium their tissues also store) and get our EPA and DHA from the same place fish get it: sea plants![412]

In January 2020, *Medical News Today* reported that "Seaweed and algae are important sources of omega-3 for people on a vegetarian or vegan diet, as they are one of the few plant groups that contain DHA and EPA." Most sushi chefs use the seaweed nori, which is an excellent source of DHA and EPA. And the algae chlorella and spirulina—readily found as supplements in the health food section of your grocery store—can be added to smoothies and oatmeal.[413]

As mentioned above, our bodies can create their own EPA and DHA from ALA, but the process isn't particularly efficient. However, we can increase our body's efficiency by reducing our consumption of omega-6 oils because these oils compete for the enzymes we need to manufacture our own EPA and DHA from ALA. According to a 2016 article in *Nutrients*, the typical Western diet is high in omega-6 fatty acids, of which sunflower, corn, soybean, and cottonseed oils are the most common sources. Lowering our consumption of omega-6s not only reduces our chance of becoming obese but also frees up more of the enzymes we need to convert ALA to EPA and DHA.[414]

You can quickly determine the EPA and DHA levels in your blood by a blood test your doctor can order; you can also get testing kits online. I recommend you include the plant-based foods listed above and reduce your omega-6 consumption and then get tested to see where your EPA and DHA levels are before spending money on supplementation.

It has never been easier to be plant-based. Sometimes it's as easy as reaching one row over to choose from a growing number of nut milks instead of cow's milk. Many groceries also carry plant-based cheese and meat substitutes. These items can help make the transition to a plant-based diet more effortless, but you can also skip them and go straight for whole grains, vegetables, fruits, legumes, nuts, and seeds. Most plant-based eaters say an entirely new world of food opened up to them when they escaped the Meatrix.

For me, being plant-based is an incredible feeling. Aligning my actions with my core values of doing as little harm as possible to the earth and animals brings me inner peace and happiness. I believe my true self began to shine more fully when I left the Meatrix. Whether it's an energetic thing, good karma, the effects of better health and more energy, or all of the above, I don't know, but I've seen how my going plant-based has planted seeds of awareness more deeply in myself and in those around me. When you become plant-based, you may experience this too. You will undoubtedly become more aware of other plant-based eaters, and you'll notice the articles and research headlines promoting plant-based eating. Plant-based eating will become part of your conscious awareness, and you will start seeing more of it in the world.

FASCINATING FACTS

- Seven nutritional deficiencies are commonly found among omnivores—four more than are found in the average plant-based eater.
- B-12 is the single deficiency commonly found in plant-based eaters that differs from their meat-eating counterparts.
- Vitamin B-12 deficiency and depletion are common in wealthier countries, particularly among the elderly, and

are most prevalent in poorer populations around the world regardless of diet.

- 96 percent of cattle in the US are supplemented with cobalt to produce vitamin B12.
- Sea microalgae and plankton, not fish, are the source of Omega 3 oils.
- 100 mg of vitamin C with a meal increases iron absorption by 67 percent.

CHAPTER 14

Making the Transition

Congratulations on starting your plant-based journey!
In this section I'll cover some things you might consider
doing when making the transition from animal-based foods to
plant-based foods. Like the previous section, you don't *need* to
do any of the following. Simply by escaping the Meatrix, you're
already vastly improving your diet. But if you want information
on how to make the transition go as smoothly as possible, this
section is for you.

A way to ease the transition from animal-based foods to
plant-based foods is to eat prebiotic foods, which improve your
gut's microbiome. Prebiotics are the fermentable, nondigestible
fiber that supports the gut, aiding digestion.[415] There are many
prebiotic vegan foods available, including chickpeas, lentils, kidney beans, tempeh, bananas, watermelon, grapefruit, bran, bar-
ley, oats, almonds, pistachios, and flax.[416]

Alternatively, or in addition to prebiotics that ferment within
the body, you can include a bit of already fermented food each
day to help with beneficial gut bacteria. Fermented plant-based
foods such as vegan yogurt, tempeh, kimchi, sauerkraut, miso,
and kombucha are full of good bacteria that can help improve
digestion and immunity.

Of course, if you prefer, you can just take a plant-based probiotic supplement as a way to assist your body's digestion. Probiotic supplements can be costly compared to prebiotic and fermented foods, but they often have much higher amounts of beneficial bacteria.

Dr. Will Bulsiewicz, the author of *Fiber Fueled*, promotes plant-based eating to restore health by enriching our gut microbiome. He says that unlike prebiotic and fermented foods that enhance the diversity of our microbiota, probiotic supplements are like an army of beneficial bacteria marching through our digestive tract. They do good while there but simply pass through our gut without actually enriching it. So remember that as a result our microbiome loses all benefits from probiotics within days of no longer taking them.[417]

Many professional athletes say their plant-based eating gives them a competitive edge, with increased energy and faster recovery times. But that might not be everyone's experience at first.

If you're like most people, then any significant change in your diet can be uncomfortable until your body adjusts. The reason some people might go through withdrawal when going plant-based is that your body must relearn how to synthesize proteins from plants. Those who have eaten the standard American diet (SAD) for their entire lives have developed an acquired dependency on animal protein. Of course, our bodies can manufacture creatine, carnitine, and choline, but when these nutrients enter our bodies preformed in animal proteins three times a day and are consumed throughout a lifetime, our bodies downregulate their own natural production.[418]

Why would our bodies do this? When the bacteria in our gut feast on the L-carnitine found in animal protein (red meat, fish, eggs, and poultry), they produce trimethylamine. Your liver converts trimethylamine into trimethylamine N-oxide (TMAO). According to the Harvard Medical School's Harvard Health Publishing, TMAO contributes to "a higher risk for both cardiovascular disease and early death from any cause."[419] In other words, your body down-regulates its production of choline in an attempt to save your life from the Meatrix!

Think of giving up animal products like giving up caffeine or cigarettes. If you've ever done that, you know you'll likely feel tired, short-tempered, and headachy until your body remembers how to function without those artificial stimulants. Of course, you can make these withdrawal symptoms disappear by drinking some caffeine or having a smoke. But if you hold off on having that cup of coffee or cigarette, then soon your body will no longer be dependent on them. Likewise, some people who give up animal products can experience similar withdrawal. If you're one of those, realize your dependency on animal protein will fade as your body upregulates its own natural production of these nutrients.

If you don't have anyone in your immediate circle supporting you, seek encouragement and support online! I have found that vegans are some of the most helpful people in the world. They love to share tips, recipes, information about finding good vegan eats, and more. Lexington has a vegan society, and I bet your city or community does as well.

Another fantastic thing about being plant-based is that you can replicate nearly any non-vegan dish and enjoy its plant-based health benefits. I can't begin to tell you the number of recipes I've veganized over the years. You can easily find tons of recipes online, and once you've done a few, you'll find it easy to convert the unhealthy comfort foods from your childhood—such as macaroni and cheese, lasagna, and pizza—into delicious, healthier, plant-based versions.

When you escape the Meatrix and start reaping the benefits of eating only plants, you'll likely begin to notice some improvements to your health. Here's a quick story from my personal experience. As a child and adult, I suffered from seasonal allergies. As an adult, I underwent immunotherapy—allergy shots. After years of injections, the treatment was successful, and I no longer required the regimen. However, I still saw my allergist every six months for follow-ups. During one of these checkups, my doctor noticed a pattern. I would come to her office for antibiotics after every international trip. But a delightful and unexpected thing happened when I went from vegetarian to 100 percent plant-based. I no longer visited her office after my international travel!

Not needing antibiotics is excellent news because they kill both the good and bad bacteria in our gut, so it's best to avoid them whenever possible. Also, frequent use of them encourages the bacteria to develop resistance. So breaking the pattern of going on antibiotics after every international flight is what I call thriving outside the Meatrix! And I haven't needed to see my allergist in many years, which is a better use of her time and mine, plus it saves me two co-pays per year. That's money in the bank!

As mentioned in Chapter 12, there's a misconception that plant-based eaters are weak and that strength and fitness require eating meat. But in reality the world now has multiple world-record-holding plant-based athletes. Given the relatively small size of the current plant-based movement, it's incredible that there would be so many highly successful plant-based athletes. Nevertheless, these are a few of the folks who are competing and winning on a plant-based diet:

- **Scott Jurek** gave up eating meat in 1997 and went 100 percent plant-based in 1999. He's won nearly all of ultra-running's elite trail and road events. His "signature race" is the Western States 100-Mile Endurance Run, which he's won a record seven times straight. In addition, in 2015 he became the world record holder for speed on the 2,190-mile-long Appalachian Trail, completing it in forty-six days, averaging fifty miles a day! He's also the author of a *New York Times* bestselling book, *Eat & Run*.
- **Rich Roll** is a triathlete, best-selling author, and host of the hugely popular *Rich Roll* podcast. In 2010, Roll accomplished a feat that many believed impossible when he completed five ironman-distance triathlons on five different Hawaiian islands in less than a week. Roll writes about his journey in his 2012 inspirational memoir, *Finding Ultra*.
- **Lewis Hamilton**, a British racing driver competing in Formula 1, became plant-based in 2017. In 2021 Hamilton bagged his one hundredth Grand Prix win, making him the first Formula 1 driver to do so.
- **Kate Strong** is the first female athlete to set a record, pedaling 433.1 miles in twenty-four hours on a static bike. Strong went plant-based in 2012 to increase lung capacity. After giving up dairy, her persistent asthma disappeared, and she's never looked back.
- **Sophia Ellis** ranked fourth in the world and took home the bronze medal for deadlift at the 2021 International Powerlifting Federation. Ellis deadlifted 227.5 kg (501.5 pounds), breaking the previous women's deadlift record. Ellis became plant-based in 2015 and currently holds fourteen national and international records.
- **Venus Williams** has been crushing it on the tennis court for years. So far in her illustrious career, she's won seven

Grand Slam singles titles, fourteen Grand Slam doubles titles, and five women's singles titles at Wimbledon—including twice as a raw-food vegan (in 2012 and 2016). She also has four Olympic gold medals. In 2011, Williams adopted a plant-based lifestyle after being diagnosed with Sjögren's syndrome. Williams now has her own line of plant-based nutritional products called Happy Viking.

- **Patrik Baboumian,** a world-record-holding strongman, became entirely plant-based in 2011 and in the same year gained the title of Germany's Strongest Man. In 2015, he set his fourth world record when he surpassed his previous world record by completing a yoke walk carrying 560 kg (1,230 pounds) in only twenty-eight seconds!
- After going vegan in 2013, surfer **Tia Blanco** took the gold medal at the 2015 International Surfing Association Open Women's World Surfing Championship in Popoyo, Nicaragua. Blanco successfully defended her title the following year in Playa Jaco, Costa Rica.
- **Kendrick Farris** became vegan in 2014 and was the only male weightlifter from Team USA to qualify for the 2016 Olympic Games in Rio. He moved up a weight class on a plant-based diet and lifted 831 pounds at the US Olympic Trial, setting a new American record.
- **Nate Diaz** is a thirty-six-year-old Ultimate Fighting Champion (UFC) and has been plant-based since he was eighteen years old. Diaz holds an unprecedented eight "Fight of the Night" awards.
- **Meghan Duhamel** is a Canadian figure skater who has been vegan since 2008. At the 2018 Winter Olympics, at thirty-two, she and her skating partner Eric Radford won a bronze medal and contributed to Canada's team gold. In 2018 Duhamel and Radford made history by complet-

ing the first-ever quadruple throw jump at any Winter Olympics.

And the list goes on and on. Plant-based athletes have been competing and winning at very high levels in every imaginable sport, dispelling the notion of vegans being frail, sick, and unhealthy people. On the contrary, almost all the vegan athletes I've researched say they feel better and recover more quickly, crediting eating plants with giving them a competitive edge over meat eaters.

I've seen successes in my own life too. I didn't take up running until 2015, when I was in my fifties, seven years after going plant-based; even then, it was for fun, not competition. During this time, I've enjoyed running but never entered a race until November 24, 2019, when, at the age of fifty-seven, I surprised my friends and family by successfully running a half-marathon. That's a big success for me—an average, non-athletic man. My goal was to do another half-marathon the following year, but the coronavirus pandemic threw a wrench in those plans. But in 2021, I decided to see if I could run the distance of a half-marathon (13.1 miles) more quickly than my half-marathon race in 2019. After training for six weeks, I ran the 13.1 miles twenty-seven minutes and thirty-four seconds faster than my 2019 half-marathon.

When plant-based Olympic athletes shatter records, it's easy to chalk that up to their superior athleticism. But it's not just athletes who are powered by plants. People from all walks of life—such as musicians, actors, doctors, mental health professionals, lawyers, and accountants—are also feeling great outside the Meatrix.

Finally...Highlighting the Basics

Thank you for going on this journey with me. I hope you can now see the many ways the dystopian world described in the introduction of this book, as far-fetched as it may have sounded then, is real, and we're currently living in it. We each have the ability, and I'd say an obligation, to disrupt the powers creating a hellish future for our world. The great news is we don't need to buy an expensive electric car. We don't need to wait for the cost of solar panels to become affordable. We don't need to suffer from low energy and poor health. We don't need to contribute to animal suffering. We don't need to wait—and shouldn't. We need to act *now*!

Global governments and corporations collude with the Meatrix to keep us addicted to their unhealthy and polluting products, so it is a fool's folly waiting on them to change. Instead, we can start lowering our carbon footprint by 73 percent today and kicking climate change's ass by escaping the Meatrix. The best news is it's never been easier or more delicious.

In doing research for this book, I was amazed at the proliferation of scientific data supporting eating plants. For example, climate change is absolutely real, and science shows that there are things we can do to bring about change—a whole-foods, plant-based lifestyle is not only healthier for you, but it's also better for the planet and the animals.

According to the United Nations, emerging infectious diseases and climate change are the two greatest threats facing humanity and our planet. The coronavirus pandemic is far from over, and the most recent scientific data reveals that those following a plant-based dietary pattern (even without vaccination) have an astonishing 73 percent lower chance of contracting moderate to severe COVID-19. So not only will eating only plants offer you protection against this pandemic, but it could also help when the next one emerges. But more importantly, eating a plant-based diet lowers the likelihood of future pandemics by significantly reducing our incursion into natural habitats where previously unknown viruses may be lurking.

Sadly, climate change is upon us. For example, the National Oceanic Atmospheric Association (NOAA) says the average temperature in mainland US was 2.5 degrees Fahrenheit above the twentieth-century average. Further, "The six warmest years on record have all occurred since 2012."[420]

Fortunately, some leaders finally recognize the need for us to divest ourselves from fossil fuels. A few corporations are committing to reduce their carbon footprints, with too few pledging to become completely carbon neutral in the decades to come. The transition from gas-powered vehicles to electric cars is on the horizon. Yet retooling manufacturing and building new infrastructure takes time. It will likely be many years before we see any meaningful reductions in greenhouse gases due to these measures.

Since the most recent UN study on climate change has come out, people worldwide have been looking at ways to make a difference for the planet. While there are many small ways to lower your carbon footprint, only one thing will reduce it by an incredible 73 percent *today*, and that is to escape the Meatrix.

The number one cause of death in the US is heart disease. A plant-based lifestyle is scientifically proven to reverse heart disease with documented results in as little as three weeks! So shouldn't a dietary pattern proven to prevent America's number one cause of death become the de facto pattern? When you add the prevention of heart disease to the litany of other chronic diseases plaguing Western societies that improve or disappear when eating only plants, it just makes sense to forgo animal-based foods and join the plant-based revolution.

According to a survey by the American Society for the Prevention of Cruelty to Animals, 94 percent of the public agrees that "from every step of their lives on a farm—from birth to slaughter—farm animals should be treated in a way that inflicts the least amount of pain and suffering possible."[421] The Meatrix does immeasurable harm to animals, and we're complicit in that suffering any time we purchase animal-based foods. Speciesism based on an arbitrary and inaccurate hierarchy fuels the Meatrix, and flawed thinking leads to greater suffering for all human and nonhuman animals.

Growing crops to feed animals increases deforestation, fuels climate change, increases the likelihood of newly emerging infectious diseases, and destroys rich biospheres, replacing them with monocultures that lead to increased extinction rates and land degradation. Growing crops to feed humans instead of livestock is a much better land-management technique and offers a greater return on investment. The food the US feeds to animals alone could feed one billion people. Globally, the number is four billion people—more than half of the world's population! For those concerned about malnutrition worldwide, a plant-based diet revolution could redirect food to reach those humans needing it the most.

In the final scene of *The Matrix Resurrections*, the Analyst (a psychiatrist played by Neil Patrick Harris) informs newly resurrected and empowered Trinity and Neo that "the sheeple aren't going anywhere. They like my world. They don't want this sentimentality. They don't want freedom or empowerment. They want to be controlled. They crave the comfort of certainty. And that means you two back in your pods. Unconscious and alone. Just like them."

Neo and Trinity's reply is they just stopped by to say thanks for giving them a chance to build a new world.

How about you? Do you want to build a new and better world, or are you one of the people who the Analyst says are going nowhere? Do you want to be controlled by the meat–industrial complex? Do you prefer the comfort and certainty of the Meatrix despite now knowing the costs to your health, the health of our planet, and its animals? Do you want to be unconscious and alone like the humans in *The Matrix*?

ACKNOWLEDGMENTS

Thanks to all my friends for believing in me and the message of my book.

Special thanks to Samuel, without whom this book would not be possible. Thanks are not enough to my best friend, life partner, and fierce editor, Lea. To all those who helped along the way with encouragement and advice, particularly David, Frank, and Mary Claire. Much gratitude to Chris, Dale, Karen, Lakshmi, Nachie, Rebecca, and Saeeda, who helped improve this book.

Enormous gratitude to those who helped me further offset the publishing costs of the book by becoming Titanium Level contributors in my FundRazr campaign. They are:

- Anonymous
- River Birch Publishing LLC
- Lea and Frank Schultz
- Lakshmi Sriraman

And to everyone at Scribe Media, especially AJ, Aleza, Chelsea, Erin, Esty, Holly, Geoff, Joy, Mark, Natalia, and Sky. This book wouldn't have happened without your help. I am so grateful.

The Five Fundamentals for a Successful Transition to a Whole-Foods, Plant-Based Lifestyle

I reached out to Bruce Garry, a nutritional advisor certified by the T. Colin Campbell Center for Nutrition Studies at eCornell and a certified CHIP (Complete Health Improvement Program) facilitator who specializes in advising people who are embarking on a whole-foods, plant-based lifestyle. As scientific research now recognizes (see the Introduction), Mr. Garry says, "Your genes are not your destiny, and gene expression is heavily influenced by lifestyle." Additionally, he says that by adopting a whole-foods, plant-based lifestyle, we're embarking on a course that's on par with people living in the famed blue zones of the world—the hot spots for longevity and high concentrations of people who avoid many of Western society's chronic illnesses and cancers. Mr. Garry continued, "The great news is you can reap many of the same benefits of blue zones without moving to Japan or Greece!"

Concerning what plant-based eaters can do to set themselves up for success, Mr. Garry shared his five fundamentals for success on a plant-based lifestyle, which, he says, is key to keeping you on your path to a healthy lifestyle and a rich and fulfilling life.

1. Set Goals for Success—Your Genes Are Not Your Destiny

- Understand your personal health needs and wants.
- Know why you are making changes, be it getting healthier, living longer, overcoming chronic disease, reversing diabetes, losing weight, reducing blood pressure, etc.
- Know that how your genes are expressed is influenced by your lifestyle.
- Learn about family chronic disease history (parents, grandparents, siblings).

2. Create Your Team: Find Like-Minded People

- Create or find a community of like-minded people for support and continuing education, including family, partners, friends, doctors, etc.
- Use social media, local groups, and communities to support your efforts.
- Use internet resources to check your facts: pubmed.gov, Nutritionfacts.org, PCRM.org, forksoverknives.com, shaneandsimple.com, nutmegnotebook.com, etc.
- Create your personal healthcare team by keeping your doctors informed of your whole-foods, plant-based lifestyle.

3. Clean Out and Restock Your Pantry to Create a Plant-Based Kitchen

- Remove anything that is not on your eating plan.
- The food that's in your house ends up in your mouth.
- Eat vegetables, fruits, grains, legumes, seeds, and nuts.
- Before you eat, look at your plate to see if it is colorful. Eating the rainbow will fill your plate with phytonutrients and strengthen your microbiome.

4. Be Adventurous and Creative, and Try New Recipes Weekly

- Expand your taste experiences.
- Increase your enjoyment of eating by making your meals special.
- Share foods and explore new recipes with others.
- Eat breakfast like a king, lunch like a prince, and dinner like a pauper.
- Do batch cooking once a week for meals, grains, condiments, sauces, etc.

5. This Is a Journey. Have Fun, Enjoy New Recipes, and Eat to Feed and Heal Your Body

- It's all about making whole-food, plant-based eating your lifestyle.
- Your whole-foods, plant-based lifestyle changes must be rewarding to become a habit.
- If you make a mistake, get back up and start over with the next meal.
- Review your progress regularly. Are you reaching your goals? If not, why not?

- Nothing is set in stone. This is your plan and your adventure.

NOTES

1 United Health Foundation, "Annual Report 2021: Kentucky," America's Health Rankings, accessed May 17, 2022, https://www.americashealthrankings.org/explore/annual/measure/Health_Status/state/KY.

2 United Health Foundation, "Annual Report 2021: West Virginia," America's Health Rankings, accessed May 20, 2022, https://www.americashealthrankings.org/explore/annual/measure/Health_Status/state/WV.

3 Centers for Disease Control and Prevention National Center for Health Statistics, "Cancer Mortality by State," reviewed February 28, 2022, https://www.cdc.gov/nchs/pressroom/sosmap/cancer_mortality/cancer.htm.

4 Centers for Disease Control and Prevention National Center for Health Statistics, "Drug Overdose Mortality by State," reviewed March 1, 2022, https://www.cdc.gov/nchs/pressroom/sosmap/drug_poisoning_mortality/drug_poisoning.htm.

5 Centers for Disease Control and Prevention National Center for Health Statistics, "Septicemia Mortality by State," reviewed March 2, 2022, https://www.cdc.gov/nchs/pressroom/sosmap/septicemia_mortality/septicemia.htm.

6 Centers for Disease Control and Prevention National Center for Health Statistics, "Chronic Lower Respiratory Disease Mortality by State," reviewed February 28, 2022, https://www.cdc.gov/nchs/pressroom/sosmap/lung_disease_mortality/lung_disease.htm.

7 Centers for Disease Control and Prevention National Center for Health Statistics, "Heart Disease Mortality by State," reviewed February 25, 2022, https://www.cdc.gov/nchs/pressroom/sosmap/heart_disease_mortality/heart_disease.htm.

8 Centers for Disease Control and Prevention National Center for Health Statistics, "Kidney Disease Mortality by State," reviewed March 1, 2022, https://www.cdc.gov/nchs/pressroom/sosmap/kidney_disease_mortality/kidney_disease.htm.

9 Sharon H. Bergquist, "Eat for Your Genes: Why a Good

Diet Matters More than Bad Genes," DrSharonBergquist.
com, July 18, 2015, https://drsharonbergquist.com/2015/07/
eat-for-your-genes-why-a-good-diet-matters-more-than-bad-genes/.

10 United States Department of Agriculture, National Agricultural Statistics Service, *2017 Census of Agriculture: United States Summary and State
Data*, AC-17-A-51, 2019, https://www.nass.usda.gov/Publications/AgCen-
sus/2017/index.php#full_report.

11 Alvin Chang et al., "The Pandemic Exposed the Human
Cost of the Meatpacking Industry's Power: 'It's Enor-
mously Frightening'," *The Guardian*, November 16, 2021,
https://www.theguardian.com/environment/2021/nov/16/
meatpacking-industry-covid-outbreaks-workers.

12 Mike Dorning, "Meatpackers Ignored Covid Spread to Keep
Operating, House Report Says," Bloomberg, May 12, 2022,
https://www.bloomberg.com/news/articles/2022-05-12/
meatpackers-ignored-covid-spread-to-keep-operating-report-says.

13 Chang et al., "Pandemic Exposed the Human Cost."

14 Chang et al., "Pandemic Exposed the Human Cost."

15 Chang et al., "Pandemic Exposed the Human Cost."

16 Chang et al., "Pandemic Exposed the Human Cost."

17 Marta Zaraska, *Meathooked: The History and the Science of Our 2.5-Million-
Year Obsession with Meat* (New York: Basic Books, 2016), 96.

18 Zaraska, *Meathooked*, 96.

19 John Dunham & Associates, "The Meat and Poultry Industry Creates Jobs in the United States," North American Meat
Institute, 2020, https://nami.guerrillaeconomics.net/reports/
be7cd451-8c70-4265-8867-e8adc3cc4a9b.

20 Felix Richter, "Americans Develop a Taste for Dairy and Meat Alternatives," Statista, July 23, 2021, https://www.statista.com/chart/25390/
sales-of-plant-based-meat-and-dairy-alternatives/.

21 "USDA Strategic Goals, 2018–2022," United States Department of Agriculture, May 2018, https://www.usda.gov/sites/default/files/documents/
usda-strategic-goals-2018-updated-1.pdf.

22 Karl Michaëlsson et al., "Milk Intake and Risk of Mortality and Fractures
in Women and Men: Cohort Studies," *The BMJ* 349 (2014): g6015, https://
doi.org/10.1136/bmj.g6015.

23 Michaëlsson et al., "Milk Intake and Risk of Mortality."

24 Diane Feskanich et al., "Milk Consumption During Teenage Years and
Risk of Hip Fracutres in Older Adults," *JAMA Pediatrics* 168, no. 1 (2014):
54–60, https://doi.org/10.1001/jamapediatrics.2013.3821.

25 NIH Osteoporosis and Related Bone Disease National Resource

Center, "Exercise for Your Bone Health," reviewed October 2018, https://www.bones.nih.gov/health-info/bone/bone-health/exercise/exercise-your-bone-health.

26 NHS, "Food for Healthy Bones," reviewed April 8, 2021, https://www.nhs.uk/Live-well/bone-health/food-for-strong-bones.

27 IPSOS Retail Performance, "Vegan Trends in the U.S.," accessed January 25, 2022, https://www.ipsos-retailperformance.com/en/vegan-trends.

28 Nils-Gerrit Wunsch, "Number of Vegans in Great Britain 2014–2019," Statista, June 22, 2021, https://www.statista.com/statistics/1062104/number-of-vegans-in-great-britain.

29 Quoted in Lucy Danziger, "The Number of Americans Eating Plant-Based Has Passed 9.7 Million," The Beet, March 9, 2020, https://thebeet.com/the-number-of-americans-eating-plant-based-has-passed-9-7-million-survey-finds.

30 Nathaniel Popper, "Beyond Meat's Share Price Surges on First Day of Trading," *The New York Times,* May 2, 2019, https://www.nytimes.com/2019/05/02/technology/beyond-meat-ipo-stock-price.html.

31 Mikaela Cohen, "Impossible Foods, Beyond Meat Battle to Achieve Price Parity with Real Meat," CNBC Evolve, August 25, 2021, https://www.cnbc.com/2021/08/25/impossible-foods-beyond-meat-battle-price-parity-with-real-meat.html.

32 Mary Meisenzahl, "Burger King Doubles Down with Impossible Nuggets after Impossible Whopper Success," *Insider*, October 6, 2021, https://www.businessinsider.com/burger-king-will-sell-impossible-nuggets-after-impossible-whopper-success-2021-10.

33 Quoted in Jemima Webber, "Burger King's Flagship London Restaurant Is Going Fully Vegan for a 1-Month Trial," Plant Based News, March 10, 2022, https://plantbasednews.org/lifestyle/food/burger-king-vegan-london.

34 Quoted in Helena Horton, "Burger King Ends All-Vegan London Branch Trial Amid Prediction Trend Will Become Norm," *The Guardian*, April 14, 2022, https://www.theguardian.com/lifeandstyle/2022/apr/14/meat-feasts-to-go-burger-king-tests-all-vegan-london-branch.

35 Renub Research, *United States Plant Based Food Market Forecast by Segments, Food Services, Merger and Acquisitions, Company Analysis* (Roswell, GA: Renub Research, 2021), https://www.renub.com/united-states-plant-based-food-market-nd.php.

36 Hannah Ritchie and Max Roser, "Meat and Dairy Production," Our World in Data, August 2017, https://ourworldindata.org/meat-production#citation.

37 Timothy Robinson and Francesca Pozzi, *Mapping Supply and Demand for*

Animal-Source Foods to 2030 (Rome: Food and Agriculture Organization of the United Nations, 2011), https://www.fao.org/3/i2425e/i2425e00.pdf.

38 Joseph Poore, "The Impact of Meat and Dairy on the Planet," Million Dollar Vegan, February 10, 2019, video, 5:08, https://www.youtube.com/watch?v=WJ25O_BD5OM.

39 Beth Mole, "Big Phrama Shells Out $20B Each Year to Schmooze Docs, $6B on Drug Ads," Ars Technica, January 11, 2019, https://arstechnica.com/science/2019/01/healthcare-industry-spends-30b-on-marketing-most-of-it-goes-to-doctors.

40 American Society for the Prevention of Cruelty to Animals (ASPCA), "ASPCA Research Shows Americans Overwhelmingly Support Investigations to Expose Animal Abuse on Industrial Farms," press release, February 17, 2012, https://www.aspca.org/about-us/press-releases/aspca-research-shows-americans-overwhelmingly-support-investigations-expose.

41 American Society for the Prevention of Cruelty to Animals, (ASPCA) "ASPCA Surveys," accessed April 8, 2022, https://www.aspca.org/shopwithyourheart/business-and-farmer-resources/aspca-surveys.

42 American Society for the Prevention of Cruelty to Animals (ASPCA), "What Is Ag-Gag Legislation?" accessed January 26, 2022, https://www.aspca.org/improving-laws-animals/public-policy/what-ag-gag-legislation.

43 ASPCA, "What Is Ag-Gag."

44 John A. Kehoe, "The Story of Biosynthetic Human Insulin," in *Frontiers in Bioprocessing*, ed. Subhas K. Sikdar, Milan Bier, and Paul Todd (Boca Raton: CRC Press, 1989), 45–49.

45 Equine Wellness, "Premarin Horses," *Equine Wellness Magazine*, January 14, 2014, https://equinewellnessmagazine.com/premarin-horses.

46 Certified Humane, "'Free Range' and 'Pasture Raised' Oficially Defined by HFAC for Certified Humane® Label," CertifiedHumane.org, January 16, 2014, https://certifiedhumane.org/free-range-and-pasture-raised-officially-defined-by-hfac-for-certified-humane-label.

47 People for the Ethical Treatment of Animals (PETA), "Animals Used for Free-Range and Organic Meat," accessed January 26, 2022, https://www.peta.org/issues/animals-used-for-food/organic-free-range-meat.

48 Will Bulsiewicz, *Fiber Fueled: The Plant-Based Gut Health Program for Losing Weight, Restoring Your Health, and Optimizing Your Microbiome* (New York: Avery, 2020), xvii–xviii.

49 Jessie Szalay, "What Are Phytonutrients?" LiveScience, October 21, 2015, https://www.livescience.com/52541-phytonutrients.html.

50 Mayo Clinic Staff, "Dietary Fiber: Essential for a Healthy Diet," Mayo Clinic, January 6, 2021, https://www.mayoclinic.org/healthy-lifestyle/

nutrition-and-healthy-eating/in-depth/fiber/art-20043983.

51 Marta Guasch-Ferré and Geng Zong, "Mono-Unsaturated Fats from Plants, Not Animals May Reduce Risk of Death from Heart Disease and Other Causes," meeting report, American Heart Association, March 21, 2018, https://newsroom.heart.org/news/mono-unsaturated-fats-from-plants-not-animals-may-reduce-risk-of-death-from-heart-disease-and-other-causes.

52 Guasch-Ferré and Zong, "Mono-Unsaturated Fats."

53 Kimberly Holland, "What Are the 12 Leading Causes of Death in the United States?" Healthline, updated March 29, 2019, https://www.health-line.com/health/leading-causes-of-death.

54 Lisa Hark and Darwin Deen, "Taking a Nutrition History: A Practical Approach for Family Physicians," *American Family Physician* 59, no. 6 (March 1999): 1521, https://www.aafp.org/afp/1999/0315/p1521.html; Courtland Milloy, "Training Doctors to Talk about the Link Between Food and Health Could Be the Converation We Need to Save Lives," *The Washington Post*, December 3, 2019, https://www.washingtonpost.com/local/training-doctors-to-talk-about-the-link-between-food-and-health-could-be-the-conversation-we-need-to-save-lives/2019/12/03/d91897f4-15f8-11ea-9110-3b34ce1d92b1_story.html.

55 Stacey Colino, "How Much Do Doctors Learn About Nutrition?" *U.S. News & World Report*, December 7, 2016, https://health.usnews.com/wellness/food/articles/2016-12-07/how-much-do-doctors-learn-about-nutrition.

56 Stephen Devries and Andrew M. Freeman, "Nutrition Education for Cardiologists: The Time Has Come," *Current Cardiology Reports* 19, no. 77 (2017), https://doi.org/10.1007/s11886-017-0890-6.

57 Bulsiewicz, *Fiber Fueled*, xix.

58 Johnston et al., "Unprocessed Red Meat and Processed Meat Consumption: Dietary Guideline Recommendations from the Nutritional Recommendations (NutriRECS) Consortium," *Annals of Internal Medicine* 171, no. 10 (2019): 756–764, https://doi.org/10.7326/M19-1621.

59 Neal Barnard, "Journal Advice to Eat Cancer-Causing Meats: Science or Clickbait?" Physicians Committee for Responsible Medicine, September 30, 2019, https://www.pcrm.org/news/blog/journal-advice-eat-cancer-causing-meats-science-or-clickbait.

60 Debbie Koenig, "Controversial Studies Say It's OK to Eat Red Meat," Nourish by WebMD, September 30, 2019, updated October 7, 2019, https://www.webmd.com/diet/news/20190930/controversial-studies-say-its-ok-to-eat-red-meat.

61 Renger F. Witkamp et al., "Let Thy Food Be Thy Medicine...When Pos-

sible," *European Journal of Pharmacology* 836 (2018): 102–114, https://doi.org/10.1016/j.ejphar.2018.06.026.

62   Caldwell B. Esselstyn et al., "A Way to Reverse CAD?" *The Journal of Family Practice* 63, no. 7 (2014): 356–364b, https://dresselstyn.com/JFP_06307_Article1.pdf.

63   Esselstyn et al., "A Way to Reverse CAD?"

64   Dean Ornish et al., "Intensive Lifestyle Changes for Reversal of Coronary Heart Disease," *JAMA* 280, no. 23 (1998): 2001–2007, https://www.doi:10.1001/jama.280.23.2001.

65   Ornish et al., "Intensive Lifestyle Changes."

66   Shamard Charles, "Too Many People Stop their Lifesaving Statins, Doctors Say," NBC News, August 6, 2017, https://www.nbcnews.com/health/heart-health/too-many-people-stop-their-lifesaving-statins-doctors-say-n789686.

67   Taylor F. Huffman et al., "Statins for the Primary Prevention of Cardiovascular Disease," Cochrane, January 31, 2013, https://www.cochrane.org/CD004816/VASC_statins-primary-prevention-cardiovascular-disease.

68   "Home," NNT.com, accessed January 28, 2022, https://www.thennt.com.

69   David Newman, "Statins Given for 5 Years for Heart Disease Prevention (With Known Heart Disease)," NNT.com, November 2, 2013, https://www.thennt.com/nnt/statins-for-heart-disease-prevention-with-known-heart-disease.

70   Esselstyn et al., "A Way to Reverse CAD?"

71   Shamard Charles, "Too Many People Stop Their Lifesaving Statins, Dotors Say," *NBC News* online, August 6, 2017, https://www.nbcnews.com/health/heart-health/too-many-people-stop-their-lifesaving-statins-doctors-say-n789686.

72   Glynn Tonsor, Ted Shcroeder, and James Mintert, "U.S. Beef Demand Drivers and Enhancement Opportunities: A Research Summary (Four-Page KSU Fact Sheet)," AgManager.info, January 30, 2009, https://www.agmanager.info/livestock-meat/marketing-extension-bulletins/trade-and-demand/us-beef-demand-drivers-and-enhancement; Richard Waite, "2018 Will See High Meat Consumption in the U.S., but the American Diet Is Shifting," World Resources Institute, January 24, 2018, https://www.wri.org/insights/2018-will-see-high-meat-consumption-us-american-diet-shifting.

73   Nathalie Bergeron et al., "Effects of Red Meat, White Meat, and Nonmeat Protein Sources on Atherogenic Lipoprotein Measures in the Context of Low Compared with High Saturated Fat Intake: A Randomized Controlled Trial," *The American Journal of Clinical Nutrition* 110, no. 1 (2019): 24–33, https://doi.org/10.1093/ajcn/nqz035.

74 American Heart Association, "Understand Your Risks to Prevent a Heart Attack," Heart.org, updated June 30, 2016, https://www.heart.org/en/health-topics/heart-attack/understand-your-risks-to-prevent-a-heart-attack.

75 Melonie Heron and Robert N. Anderson, "Changes in the Leading Cause of Death: Recent Patterns in Heart Disease and Cancer Mortality," CDC.gov, August 2016, https://www.cdc.gov/nchs/products/databriefs/db254.htm.

76 American Heart Association, "CDC: U.S. Deaths from Heart Disease, Cancer on the Rise," Heart.org, August 4, 2016, https://www.heart.org/en/news/2018/05/01/cdc-us-deaths-from-heart-disease-cancer-on-the-rise.

77 Mayo Clinic Staff, "Nutrition and Healthy Eating," Mayo Clinic, November 19, 2021, https://www.mayoclinic.org/healthy-lifestyle/nutrition-and-healthy-eating/basics/nutrition-basics/hlv-20049477.

78 "Cancer: Carcinogenicity of the Consumption of Red Meat and Processed Meat," Q&A, World Health Organization, October 26, 2015, https://www.who.int/news-room/questions-and-answers/item/cancer-carcinogenicity-of-the-consumption-of-red-meat-and-processed-meat.

79 Diana Kwon, "Myth Busters: Does this Food Cause Cancer?" WebMD, accessed January 28, 2022, January 17, 2022, https://www.webmd.com/cancer/features/does-this-food-cause-cancer.

80 Mohammad H. Forouzanfar et al., "Global, Regional, and National Comparative Risk Assessment of 79 Behavioural, Environmental and Occupational, and Metabolic Risks or Clusters of Risks in 188 Countries, 1990–2013: A Systematic Analysis for the Global Burden of Disease Study 2013," The Lancet 386, no. 10010 (2015): P2287–2323, https://doi.org/10.1016/S0140-6736(15)00128-2.

81 "Dairy Increases Risk for Death from Prostate Cancer," Physicians Committee for Responsible Medicine, June 1, 2015, https://www.pcrm.org/news/health-nutrition/dairy-increases-risk-death-prostate-cancer.

82 Ralph Ellis, "U.S. Has Record 4,237 COVID Deaths in One Day," WebMD, January 14, 2021, https://www.webmd.com/lung/news/20210114/u-s-has-record-4327-covid-deaths-in-one-day.

83 Al Cross, "Adult Obesity in Ky. Reaches All-Time High of 36.6%, Fifth in U.S.; Doctor Says Insurance Needs to Start Covering Obesity Prevention," Kentucky Health News, September 16, 2019, https://ci.uky.edu/kentuckyhealthnews/2019/09/16/adult-obesity-in-ky-reaches-all-time-high-of-36-6-fifth-in-u-s-doctor-says-insurance-needs-to-start-covering-obesity-prevention.

84 Stefanie Eschenbacher and Adriana Barrera, "Mexico Official Calls Diet a Factor as Coronavirus Death Toll Climbs," Yahoo! News, April

4, 2020, https://www.yahoo.com/now/mexican-health-ministry-regis-ters-1-013328445.html.

85 Hyunju Kim et al., "Plant-Based Diets, Pescatarian Diets and COVID-19 Severity: A Population-Based Case–Control Study in Six Countries," *BMJ Nutrition, Prevention & Health* 4 (2021): 257–266, http://dx.doi.org/10.1136/bmjnph-2021-000272.

86 Claire Gillespie, "5 Preexisting Conditions That Can Make it Harder to Fight Coronavirus," Health.com, last modified March 17, 2020, https://www.health.com/condition/infectious-diseases/coronavirus/coronavirus-preexisting-conditions.

87 Ye Li et al., "Dietary Patterns and Depression Risk: A Meta-Analysis," *Psychiatry Research* 253 (July 2017): 373, https://doi.org/10.1016/j.psychres.2017.04.020.

88 Jen Wyglinsky, "Does Veganism Help Reduce Stress and Anxiety?" Faunalytics, May 25, 2016, https://faunalytics.org/veganism-help-reduce-stress-anxiety.

89 Bonnie Beezhold et al., "Vegans Report Less Stress and Anxiety than Omnivores," *Nutritional Neuroscience* 18, no. 7 (2015), https://doi.org/10.1179/1476830514Y.0000000164.

90 Nadia Murray-Ragg, "10 Vegan Foods to Help Reduce Feelings of Anxiety," Live Kindly, last modified April 16, 2019, https://www.livekindly.co/vegan-foods-relieve-anxiety.

91 Megu Y. Baden et al., "Quality of Plant-Based Diet and Risk of Total, Ischemic, and Hemorrhagic Stroke," *Neurology* 96, no. 15 (April 13, 2021): e1040, https://doi.org/10.1212/WNL.0000000000011713.

92 Bronwyn S. Berthon and Lisa G. Wood, "Nutrition and Respiratory Health—Feature Review," *Nutrients* 7, no. 3 (2015): 1628–1629, https://doi.org/10.3390/nu7031618.

93 Egeria Scoditti et al., "Role of Diet in Chronic Obstructive Pulmonary Disease Prevention and Treatment," *Nutrients* 11, no. 6 (2019): 21, https://doi.org/10.3390/nu11061357.

94 "Alzheimer's Disease: Boost Brain Health with a Plant-Based Diet," Physicians Committee for Responsible Medicine, accessed March 4, 2022, https://www.pcrm.org/health-topics/alzheimers.

95 Mary Chapman, "Awareness Month Activities Shine Spotlight on Alzheimer's, Dementia," Alzheimer's News Today, November 4, 2020, https://alzheimersnewstoday.com/2020/11/04/awareness-month-activities-shine-spotlight-on-alzheimers-dementia.

96 Alzheimer's Association, "Stages of Alzheimer's," Alzheimer's Association, accessed March 24, 2022, https://www.alz.org/alzheimers-dementia/stages.

97   National Institute of Health's National Institute on Aging, "What is Alzheimer's Disease," US Department of Health and Human Services, last reviewed July 8, 2021, https://www.nia.nih.gov/health/what-alzheimers-disease.

98   Alzheimer's Association, "2022 Alzheimer's Disease Facts and Figures: Special Report, more than Normal Aging: Understanding Mild Cognitive Impairment," Alzheimer's Association, 2022, https://www.alz.org/media/Documents/alzheimers-facts-and-figures.pdf.

99   Jennifer J. Manly and Richard Mayeux, "Ethnic Differences in Dementia and Alzheimer's Disease," in *Critical Perspectives on Racial and Ethnic Differences in Health in Late Life,* eds. Norman B. Anderson, Rodolfo A. Bulatao, and Barney Cohen (Washington, DC: National Academies Press, 2004).

100 Paul Giem, W.L. Beeson, and Gary E. Fraser, "The Incidence of Dementia and Intake of Animal Products: Preliminary Findings from the Adventist Health Study," *Neuroepidemiology* 12 (1993), https://doi.org/10.1159/000110296.

101 "Pillar 1: Diet & Supplements," Alzheimer's Research & Prevention Foundation, accessed March 4, 2022, https://alzheimersprevention.org/4-pillars-of-prevention/pillar-1-diet-supplements.

102 "The Truth about Fats: the Good, the Bad, and the In-Between," Harvard Health Publishing, December 11, 2019, https://www.health.harvard.edu/staying-healthy/the-truth-about-fats-bad-and-good; "Choosing Healthy Fats," Health Guide, last modified August 2021, https://www.helpguide.org/articles/healthy-eating/choosing-healthy-fats.htm.

103 Nan Hu et al., "Nutrition and the Risk of Alzheimer's Disease," BioMed Research International (2013), https://doi.org/10.1155/2013/524820.

104 Robert Preidt, "Alzheimer's-Linked Brain Plaques and Blood Flow," WebMD, November 24, 2015, https://www.webmd.com/alzheimers/news/20151124/alzheimers-linked-brain-plaques-may-also-slow-blood-flow.

105 "Pillar 1."; Ryan Raman, "12 Healthy Foods High in Antioxidants," Healthline, March 12, 2018, https://www.healthline.com/nutrition/foods-high-in-antioxidants.

106 "Pillar 1."; "Heart Disease and Homocysteine," WebMD, last reviewed August 24, 2020 by James Beckerman, https://www.webmd.com/heart-disease/guide/homocysteine-risk.

107 Anetta Undas, Jan Brożek, and Andrzej Szczeklik, "Homocysteine and Thrombosis: From Basic Science to Clinical Evidence," *Thrombosis and Haemostasis* 94, no. 5 (2005): 907–915, https://doi.org/10.1160/th05-05-0313; "Methionine—Uses, Side Effects, and More," WebMD, accessed

March 4, 2022, https://www.webmd.com/vitamins/ai/ingredientmono-42/methionine.

108 David A. Smith et al., "Homocysteine and Dementia: An International Consensus Statement," *Journal of Alzheimer's Disease* 62, no. 2 (2018): 561–570, https://doi.org/10.3233/jad-171042.

109 Michael Greger, "Reducing Glycotoxin Intake to Help Reduce Brain Loss," NutritionFacts.org, December 8, 2016, https://nutritionfacts.org/2016/12/08/reducing-glycotoxin-intake-to-help-reduce-brain-loss.

110 Chris Walling, personal communication with the author, September 26, 2021.

111 Centers for Disease Control and Prevention, "National and State Diabetes Trends," CDC.gov, last reviewed January 4, 2021, https://www.cdc.gov/diabetes/library/reports/reportcard/national-state-diabetes-trends.html.

112 Center for Disease Control and Prevention, "National Diabetes Statistics Report," CDC.gov, last reviewed January 18, 2022, https://www.cdc.gov/diabetes/data/statistics-report/index.html.

113 Kentucky Health Rankings, "Top 15 Causes of Death: Kentucky," World Life Expectancy, December 22, 2021, https://www.worldlifeexpectancy.com/top-15-causes-of-death-kentucky.

114 Michelle McMacken and Sapana Shah, "A Plant-Based Diet for the Prevention and Treatment of Type 2 Diabetes," *Journal of Geriatric Cardiology* 14, no. 5 (May 2017): 342, https://pubmed.ncbi.nlm.nih.gov/28630614.

115 Lisa G. Wood et al., "Manipulating Antioxidant Intake in Asthma: A Randomized Controlled Trial," *The American Journal of Clinical Nutrition* 96, no. 3 (September 2021), https://doi.org/10.3945/ajcn.111.032623.

116 Bronwyn S. Berthon and Lisa G. Wood, "Nutrition and Respiratory Health—Feature Review," *Nutrients* 7, no. 3 (2015): 1618-1643, https://doi.org/10.3390/nu7031618.

117 Berthon and Wood, "Nutrition and Respiratory Health."

118 Berthon and Wood, "Nutrition and Respiratory Health," 1629.

119 J.D. Edgar et al., "Increased Fruit and Vegetable Consumption Improves Antibody Response to Vaccination in Older People: The ADIT Study," *Proceedings of the Nutrition Society* 69, no. OCE3 (2010): E238, https://doi.org/10.1017/S0029665110000273.

120 Stewart D. Rose and Amanda J. Strombom, "A Plant-Based Diet Prevents and Treats Chronic Kidney Disease," *Juniper Online Journal Urology & Nephrology* 6, no. 3 (2019): 555687, https://doi.org/10.19080/JOJUN.2018.06.555687.

121 Philippe Chauveau et al., "Vegetarian Diets and Chronic Kidney Disease," *Nephrology Dialysis Transplantation* 34, no. 2 (February 2019): 199,

https://doi.org/10.1093/ndt/gfy164.

122 "Plant-Based vs. Animal-Based Diets: The Jury Is In!" National Kidney Foundation, October 17, 2019, https://www.kidney.org/newsletter/plant-based-vs-animal-based-diets-jury.

123 "Top 10 Prescription Medications in the U.S. (November 2021)," GoodRx, accessed March 7, 2022, https://www.goodrx.com/drug-guide.

124 Mayo Clinic Staff, "High Blood Pressure Dangers: Hypertension's Effects on Your Body," Mayo Clinic, January 14, 2022, https://www.mayoclinic.org/diseases-conditions/high-blood-pressure/in-depth/high-blood-pressure/art-20045868.

125 Matej Mikulic, "Number of Lisinopril Prescriptions in the U.S. from 2004 to 2019 (In Millions)," Statista, February 3, 2022, https://www.statista.com/statistics/779771/lisinopril-prescriptions-number-in-the-us.

126 "Lisinopril Side Effects by Likelihood and Severity," WebMD, accessed March 7, 2022, https://www.webmd.com/drugs/2/drug-6873-1785/lisinopril-oral/lisinopril-solution-oral/details/list-sideeffects.

127 Luke J. Laffin et al., "Rise in Blood Presusre Observed Among US Adults During the COVID-19 Pandemic," *Circulation* 145, no. 3 (January 2022), https://doi.org/10.1161/CIRCULATIONAHA.121.057075.

128 Quoted in NIH National Heart and Lung Institute, "Blood Pressure Up? COVID-19 Pandemic Could Be to Blame," Research Feature, February 1, 2022, https://www.nhlbi.nih.gov/news/2022/blood-pressure-covid-19-pandemic-could-be-blame.

129 Sarah Alexander et al., "A Plant-based Diet and Hypertension," *Journal of Geriatric Cardiology* 14, no. 5 (May 2017): 327–330, https://pubmed.ncbi.nlm.nih.gov/28630611.

130 Kimberly Holland, "Obesity Facts," Healthline, last modified January 18, 2022, https://www.healthline.com/health/obesity-facts.

131 The American Cancer Society Medical and Editorial Content Team, "American Cancer Society Guideline for Diet and Physical Activity," American Cancer Society, last modified June 9, 2020, https://www.cancer.org/healthy/eat-healthy-get-active/acs-guidelines-nutrition-physical-activity-cancer-prevention/guidelines.html.

132 Colin Mathers, Gretchen Stevens, and Maya Mascarenhas, "Global Health Risks: Mortality and Burden of Disease Attributable to Selected Major Risks," report by the World Health Organization (2009): 17, https://www.who.int/healthinfo/global_burden_disease/GlobalHealthRisks_report_full.pdf.

133 "Fast Facts: The Cost of Obesity," STOP Obesity Alliance and School of Public Health and Health Services at George Washington University, accessed March 7, 2022, https://stop.publichealth.gwu.edu/sites/stop.

publichealth.gwu.edu/files/documents/Fast%20Facts%20Cost%20of%20 Obesity.pdf.

134 Associated Press, "Water Becomes America's Favorite Drink Again," *USA Today*, March 11, 2013, https://www.usatoday.com/story/news/ nation/2013/03/11/water-americas-favorite-drink/1978959.

135 Leo Sun, "A Foolish Take: American Soda Consump- tion Plunges to a 31-Year Low," *USA Today*, August 1, 2017, https://www.usatoday.com/story/money/markets/2017/08/01/ american-soda-consumption-plunges-31year-low/103953632.

136 An excellent online BMI calculator is available at NIH National Heart Lung and Blood Institute, "Calculate Your Body Mass Index," accessed May 17, 2022, https://www.nhlbi.nih.gov/health/educational/lose_wt/ BMI/bmicalc.htm.

137 Nico S. Rizzo et al., "Nutrient Profiles of Vegetarian and Nonvegetarian Dietary Patterns," *Journal of the Academy of Nutrition and Dietetics* 113, no. 12 (December 2013): 1610–1619, https://doi.org/10.1016/j.jand.2013.06.349.

138 Esselstyn et al., "A Way to Reverse CAD?"

139 Phillip J. Tuso et al., "Nutritional Update for Physicians: Plant- Based Diets," *The Permanente Journal* 17, no. 2 (2013): 61, https://doi. org/10.7812/TPP/12-085.

140 Christopher J. Ruhm, "Excess Deaths in the United States During the First Year of COVID-19," *National Bureau of Economic Research*, no. 29503, Working Paper Series (November 2021), http://www.nber.org/ papers/w29503.

141 "United States Sees Largest Drop in Life Expectancy Since World War II," *USA Today*, July 21, 2021, https:// www.usatoday.com/videos/news/health/2021/07/21/ us-life-expectancy-sees-largest-drop-since-wwii/8038623002.

142 Centers for Disease Control and Prevention, "Life Expectancy in the U.S. Declined a Year and Half in 2020," press release, July 21, 2021, https://www.cdc.gov/nchs/pressroom/nchs_press_releases/2021/202107. htm.

143 Arielle Mitropoulos, "An American Tragedy: US COVID Death Toll Tops 700,000," ABC News, October 1, 2021, https://abcnews.go.com/Health/ american-tragedy-covid-death-toll-tops-700000/story?id=80303622%20 %E2%80%9C.

144 United Nations Environment Programme and International Livestock and Research Institute, "Preenting the Next Pandemic: Zoonotic Dis- eases and How to Break the Chain of Transmission," Nairobi, Kenya (2020), https://unsdg.un.org/sites/default/files/2020-07/UNEP-Prevent- ing-the-next-pandemic.pdf.

145 Carl Zummer and Benjamin Mueller, "New Research Points to Wuhan Market as Pandemic Origin," *The New York Times*, last modified February 27, 2022, https://www.nytimes.com/interactive/2022/02/26/science/covid-virus-wuhan-origins.html.

146 "Covid Origin: Why the Wuhan Lab-leak Theory is Being Taken Seriously," BBC News, May 27, 2021, https://www.bbc.com/news/world-asia-china-57268111.

147 Supaporn Wacharapluesadee et al., "Evidence for SARS-CoV-2 Related Coronaviruses Circulating in Bats and Pangolins in Southeast Asia," *Nature Communications* 12 (2021): 2, https://doi.org/10.1938/s41467-021-21240-1.

148 Michael Greger, "Pandemics: History and Prevention," NutritionFacts. org, created 2008, published March 27, 2020, YouTube video, 57:31, https://www.youtube.com/watch?v=7_ppXSABYLY.

149 Greger, "Pandemics: History and Prevention."

150 World Health Organization, "Zoonoses," July 29, 2020, https://www.who.int/news-room/fact-sheets/detail/zoonoses.

151 Igor V. Babkin and Irina N. Babkina, "The Origin of the Variola Virus," *Viruses* 7, no. 3 (2015): 1100–1112, https://doi.org/10.3390/v7031100.

152 Colette Flight, "Smallpox: Eradicating the Scourge," *BBC History*, last modified February 17, 2011, https://www.bbc.co.uk/history/british/empire_seapower/smallpox_01.shtml.

153 Yuki Furuse, Akira Suzuki, and Hitoshi Oshitani, "Origin of Measles Virus: Divergence from Rinderpest Virus Between the 11th and 12th Centuries," *Virology Journal* 7, no. 52 (2010), https://dx.doi.org/10.1186%2F1743-422X-7-52.

154 Greger, "Pandemics: History and Prevention."

155 World Health Organization, "Measles," December 5, 2019, https://www.who.int/news-room/fact-sheets/detail/measles.

156 Michael S. Rosenwald, "History's Deadliest Pandemics, From Ancient Rome to Modern America," *The Washington Post*, last modified October 3, 2021, https://www.washingtonpost.com/graphics/2020/local/retropolis/coronavirus-deadliest-pandemics.

157 "The Story Of...Smallpox—And Other Deadly Eurasian Germs," *Guns, Germs and Steel*, PBS, accessed March 13, 2022, http://www.pbs.org/gunsgermssteel/variables/smallpox.html.

158 Michael S. Rosenwald, "Columbus Brought Measles to the New World. It was a Disaster for Native Americans," *The Washington Post*, May 5, 2019, https://www.washingtonpost.com/history/2019/05/05/columbus-brought-measles-new-world-it-was-disaster-native-americans.

159 Maria E. Tudor, Ahmad M. Al Aboud, and William Gossman, "Syphilis,"

in *StatPearls* (Treasure Island, FL: StatPearls Publishing, 2022).

160 World Health Organization, "HIV/AIDS," November 20, 2021, https://www.who.int/news-room/fact-sheets/detail/hiv-aids.

161 Centers for Disease Control and Prevention, "Ebola (Ebola Virus Disease)," last reviewed March 21, 2022, https://www.cdc.gov/vhf/ebola/index.html.

162 Gregor, "Pandemics: History and Prevention."

163 Melvin Sanicas, "What Makes Bats the Perfect Hosts for so Many Viruses," Healthcare in America, June 28, 2018, https://healthcareinamerica.us/what-makes-bats-the-perfect-hosts-for-so-many-viruses-3274c019bb4d.

164 Yizhou Jiang et al., "Cytokine Storm in COVID-19: From Viral Infection to Immune Responses, Diagnosis and Therapy," *International Journal of Biological Sciences* 18, no. 2 (2022): 459–472, https://dx.doi.org/10.7150%2Fijbs.59272.

165 Centers for Disease Control and Prevention, "CDC SARS Response Timeline," last modified April 26, 2013, https://www.cdc.gov/about/history/sars/timeline.htm.

166 CDC, "CDC SARS Response Timeline."

167 Gloria Riviera, Nick Capote, and Lauren Effron, "Cat Poop Coffeee, the World's Most Expensive, Brews Up Animal Rights Controversy," ABC News, July 18, 2015, https://abcnews.go.com/International/civet-cat-poop-coffee-worlds-expensive-brews-animal/story?id=30011989.

168 The Mount Sinai Hospital / Mount Sinai School of Medicine, "2009 Swine Flu Pandemic Originated in Mexico, Researchers Discover," *ScienceDaily*, June 27, 2016, https://www.sciencedaily.com/releases/2016/06/160627160935.htm.

169 Simon, *Meatonomics*, 16.

170 Gavin J. D. Smith et al., "Origins and Evolutionary Genomics of the 2009 Swine-Origin H1N1 Influenza A Epidemic," *Nature* 459 (2009): 1122, https://doi.org/10.1038/nature08182.

171 Tokiko Watanabe and Yoshihiro Kawaoka, "Pathogenisis of the 1918 Pandemic Influenza Virus," *PLoS Pathogens* 7, no. 1 (2011): e1001218, https://doi.org/10.1371/journal.ppat.1001218.

172 Jorge Galindo-González, "Live Animal Markets: Identifying the Origins of Emerging Infectious Diseases," *Current Opinion in Environmental Science & Health* 25 (February 2022): 100310, https://doi.org/10.1016/j.coesh.2021.100310.

173 Michael Standaert, "Coronavirus Closures Reveal Vast Scale of China's Secretive Wildlife Farm Industry," *The Guardian*, February 24, 2020,

https://www.theguardian.com/environment/2020/feb/25/coronavirus-closures-reveal-vast-scale-of-chinas-secretive-wildlife-farm-industry.

174 "Pangolin Facts and Information," *National Geographic*, accessed March 28, 2022, https://www.nationalgeographic.com/animals/mammals/facts/pangolins.

175 Patrick Greenfield, "Ban Wildlife Markets to Avert Pandemics, Says UN Biodiversity Chief," *The Guardian*, April 6, 2020, https://www.theguardian.com/world/2020/apr/06/ban-live-animal-markets-pandemics-un-biodiversity-chief-age-of-extinction.

176 Bryona A. Jones et al., "Zoonosis Emergence Linked to Agricultural Intensification and Environmental Change," *PNAS* 110, no. 21 (May 2013): 8399, https://doi.org/10.1073/pnas.1208059110.

177 Jones et al., "Zoonosis Emergence," 8401.

178 Leah Garcés, "Reducing Pandemic Risk Begins with Ending Factory Farming," *The Hill*, April 3, 2020, https://thehill.com/opinion/energy-environment/491066-reducing-pandemic-risk-begins-with-ending-factory-farming.

179 Bernice Yeung et al., "America's Food Safety System Failed to Stop a Salmonella Epidemic. It's Still Making People Sick," *ProPublica*, October 29, 2021, https://www.propublica.org/article/salmonella-chicken-usda-food-safety.

180 "Outbreak Investigation of E. Coli: Salad Mix (December 2019)," U.S. Food & Drug Administration, May 21, 2020, https://www.fda.gov/food/outbreaks-foodborne-illness/outbreak-investigation-e-coli-salad-mix-december-2019.

181 Stacy Rapacon, "12 Iconic Restaurant Chains: Where Are They Now?" AARP, July 8, 2020, https://www.aarp.org/politics-society/history/info-2020/iconic-restaurants-closed.html; Maria Scinto, "Why You Don't See Chi-Chi's Restaurant in America Anymore," Mashed, updated Februrary 15, 2022, https://www.mashed.com/159423/why-you-dont-see-chi-chis-restaurants-in-america-anymore.

182 Yeung et al., "America's Food Safety System Failed."

183 Jones et al., "Zoonosis Emergence," 8401.

184 Vandana Shiva, *Stolen Harvest: The Hijacking of the Global Food Supply* (South End Press, 2016), 70–71.

185 Philip Oltermann, "Berlin's University Canteens Go Almost Meat-Free as Students Prioritise Climate," *The Guardian*, August 31, 2021, https://www.theguardian.com/world/2021/aug/31/berlins-university-canteens-go-almost-meat-free-as-students-prioritise-climate.

186 Georgina Gustin, "Big Meat and Dairy Companies Have Spent Millions Lobbying Against Climate Action, a New Study Finds," Inside Climate

News, April 2, 2021, https://insideclimatenews.org/news/02042021/
meat-dairy-lobby-climate-action.

187 Gustin, "Big Meat and Dairy."

188 Joseph Poore and Thomas Nemecek, "Reducing Food's Environmental Impacts Through Producers and Consumers," *Science* 360,
no. 6392 (February 2019), https://doi.org/10.1126/science.aaq0216;
Michael Pellman Rowland, "The Most Effective Way to Save
the Planet," *Forbes*, June 12, 2018, https://www.forbes.com/sites/
michaelpellmanrowland/2018/06/12/save-the-planet.

189 Pellman Rowland, "Most Effective Way to Save."

190 P.J. Gerber et al., "Tackling Climate Change Through Livestock: A Global
Assessment of Emissions and Mitigation Opportunities," Food and
Agriculture Organization of the United Nations (2013): xii, https://www.
fao.org/3/i3437e/i3437e.pdf.

191 "Livestock's Long Shadow: Environmental Issues and Options," The
Livestock, Environment, and Development Initiative, Food and Agriculture Organization of the United Nations (2006), https://www.fao.org/
docrep/010/a0701e/a0701e.pdf.

192 Jeff McMahon, "Meat and Agriculture are Worse for the Climate
than Power Generation, Steven Chu Says," *Forbes*, April 4, 2019,
https://www.forbes.com/sites/jeffmcmahon/2019/04/04/meat-and-
agriculture-are-worse-for-the-climate-than-dirty-energy-steven-chu-
says/?sh=5ffb14e11f9d.

193 Quoted in Damian Carrington, "Avoiding Meat and Dairy Is 'Single Biggest Way' to Reduce Your Impact on Earth," *The Guardian*, May 31, 2018,
https://www.theguardian.com/environment/2018/may/31/avoiding-meat-
and-dairy-is-single-biggest-way-to-reduce-your-impact-on-earth.

194 "Overview of Greenhouse Gases," United States Environmental Protection Agency, last modified November 2021, https://www.epa.gov/
ghgemissions/overview-greenhouse-gases.

195 Max Mossler, "A Closer Look at the Environmental Costs of Food,"
Sustainable Fisheries, June 11, 2018, https://sustainablefisheries-uw.org/
environmental-costs-of-food.

196 Barclay Ballard, "Fish Farming is on the Rise, But There's
an Environmental Catch," European CEO, October 18,
2019, https://www.europeanceo.com/industry-outlook/
fish-farming-is-on-the-rise-but-theres-an-environmental-catch.

197 Quoted in Carrington, "Avoiding Meat and Dairy Is 'Single Biggest Way'
to Reduce Your Impact on Earth."

198 Jacquelyn Turner, "Grass-fed Cows Won't Save the Climate, Report
Finds," *Science*, October 2, 2017, https://www.science.org/content/article/

grass-fed-cows-won-t-save-climate-report-finds.

199 Joseph Poore and T. Nemecek, "Reducing Food's Environmental Impact Through Producers and Consumers," *Science* 360, no. 6392 (June 2018), https://doi.org/10.1126/science.aaq0216.

200 "Why Carbon Markets Won't Work for Agriculture," Institute for Agriculture & Trade Policy, February 4, 2020, https://www.iatp.org/documents/why-carbon-markets-wont-work-agriculture.

201 Robert Goodland and Jeff Anhang, "Livestock and Climate Change," *World Watch* (November/December 2009): 2, https://awellfedworld.org/wp-content/uploads/Livestock-Climate-Change-Anhang-Goodland.pdf.

202 Marco Springmann et al., "Analysis and Valuation of the Health and Climate Change Cobenefits of Dietary Change," *PNAS* 113, no. 15 (March 2016), https://doi.org/10.1073/pnas.1523119113.

203 Quoted in "Plant-Based Diets Could Save Millions of Lives and Dramatically Cut Greenhouse Gas Emissions," Oxford Martin School, March 21, 2016, https://www.oxfordmartin.ox.ac.uk/news/201603-plant-based-diets.

204 Johns Hopkins Coronavirus Resource Center, accessed December 29, 2021, https://coronavirus.jhu.edu.

205 "Food Choices and the Planet," Earth Save, accessed March 7, 2022, https://www.earthsave.org/environment.htm.

206 "US Could Feed 800 Million People with Grain that Livestock Eat, Cornell Ecologist Advises Animal Scientists," *Cornell Chronicle*, August 7, 1997, https://news.cornell.edu/stories/1997/08/us-could-feed-800-million-people-grain-livestock-eat.

207 Emily S. Cassidy et al., "Redefining Agricultural Yields: From Tonnes to People Nourished Per Hactare," *Environmental Research Letters* 8, no. 3 (August 2013):6, https://iopscience.iop.org/article/10.1088/1748-9326/8/3/034015/meta.

208 Vaclav Smil, "Eating Meat: Evolution, Patterns, and Consequences," *Population and Development Review* 28, no. 4 (January 2004), https://doi.org/10.1111/j.1728-4457.2002.00599.x.

209 Hannah Ritchie, "If the World Adopted a Plant-Based Diet We Would Reduce Global Agricultural Land Use from 4 to 1 Billion Hectares," Our World in Data, March 4, 2021, https://ourworldindata.org/land-use-diets.

210 Ritchie, "If the World Adopted a Plant-Based Diet."

211 Yinon M. Bar-On, Rob Phillips, and Ron Milo, "The Biomass Distribution on Earth," *PNAS* 115, no. 25 (May 2018), https://doi.org/10.1073/pnas.1711842115.

212 Bar-On, Phillips, and Milo, "Biomass Distribution on Earth."

213 Niles Eldredge, "The Sixth Extinction," ActionBioscience.org, June 2001,

https://www.biologicaldiversity.org/programs/population_and_sustainability/extinction/pdfs/Eldridge-6th-extinction.pdf.

214 United Nations Intergovernmental Science-Policy Platform on Biodiversity and Ecosystem Services (IPBES), "Media Release: Nature's Dangerous Decline 'Unprecedented'; Species Extinction Rates 'Accelerating'," 2019, accessed March 7, 2022, https://ipbes.net/news/Media-Release-Global-Assessment.

215 Eldredge, "The Sixth Extinction."

216 Hannah Ritchie and Max Roser, "Extinctions," Our World in Data, 2021, https://ourworldindata.org/extinctions.

217 Walter Willett et al., "Food in the Anthropocene: the EAT–*Lancet* Commission on Healthy Diets from Sustainable Food Systems," *The Lancet* 393, no. 10170 (February 2019), https://doi.org/10.1016/S0140-6736(18)31788-4.

218 Willett et al., "Food in the Anthropocene."

219 Development Initiatives, 2018 Global Nutrition Report: Shining a Light to Spur Action on Nutrition, 2018, https://reliefweb.int/sites/reliefweb.int/files/resources/2018_Global_Nutrition_Report.pdf.

220 Willett et al., "Food in the Anthropocene."

221 "Feed-to-Meat Conversion Inefficiency Ratios," A Well-Fed World, last modified October 26, 2015, https://awellfedworld.org/feed-ratios.

222 Chelsea Yant, "Hungry American and the Over Production of Food," *Medium*, October 28, 2018, https://medium.com/@cyant/hungry-america-and-the-over-production-of-food-937e88b91afd.

223 Celso H. L. Silva Jr. et al., "The Brazilian Amazon Deforestation Rate in 2020 is the Greatest of the Decade," *Nature Ecology & Evolution* 5 (2021), https://doi.org/10.1038/s41559-020-01368-x.

224 Matthew Green, "Fertiliser Use is Feulling Climate-Warming Nitrogen Oxide Emissions: Study," *Reuters*, October 7, 2020, https://www.reuters.com/article/us-climate-change-no2/fertiliser-use-is-fuelling-climate-warming-nitrous-oxide-emissions-study-idUSKBN26S35W.

225 Willett et al., "Food in the Anthropocene."

226 Arjen Y. Hoekstra, "The Water Footprint of Food," *Water for Food* (Stockholm: The Swedisch Research Council for Environment, Agricultural Sciences and Spatial Planning, 2008), 53, https://www.waterfootprint.org/media/downloads/Hoekstra-2008-WaterfootprintFood.pdf.

227 Arjen Y. Hoekstra and Mesfin M. Mekonnen, "The Water Footprint of Humanity," *PNAS* 109, no. 9 (February 2012), https://doi.org/10.1073/pnas.1109936109.

228 Hoekstra and Mekonnen, "The Water Footprint of Humanity."

229 Luis Villazon, "Which Vegan Milk is Best for the Environment," *BBC Sci-*

ence Focus Magazine, accessed March 10, 2022, https://www.sciencefocus.com/science/which-vegan-milk-is-best-for-the-environment.

230 Chelsea Harvey, "We Are Killing the Environment One Hamburger at a Time," *Business Insider*, March 5, 2015, https://www.businessinsider.com/one-hamburger-environment-resources-2015-2.

231 Martin C. Heller and Gregory A. Keoleian, "Beyond Meat's Beyond Burger Life Cycle Assessment: A Detailed Comparison Between a Plant-Based and an Animal-Based Protein Source," University of Michigan Center for Sustainable Systems, September 14, 2018, https://css.umich.edu/publication/beyond-meats-beyond-burger-life-cycle-assessment-detailed-comparison-between-plant-based.

232 Felix Richter, "Ida Smashes Previous Rainfall Records for New York City," Statista, September 3, 2021, https://www.statista.com/chart/25690/new-york-city-hourly-rainfall-records.

233 Matt Troutman, "At Least 12 Dead After Hurricane Ida Remnants Flood NYC," *Patch*, September 2, 2021, https://patch.com/new-york/new-york-city/least-8-dead-after-hurricane-ida-remnants-flood-nyc.

234 Chris Dolce, "Summer 2021 Was Hottest on Record in the Contiguous U.S., NOAA Says," *The Weather Channel* online, September 9, 2021, https://weather.com/news/climate/news/2021-09-09-summer-hottest-on-record-united-states-noaa.

235 "Mammals," National Park Service, February 3, 2020, https://www.nps.gov/fopu/learn/nature/mammals.htm.

236 Hussain Kanchwala, "How Did We Start Drinking Milk of Ruminants? Are We the Only Species to Drink Milk of Other Species?" ScienceABC, March 28, 2022, https://www.scienceabc.com/humans/species-drink-milk-another-species.html.

237 Pete Bauman and Allen Williams, "Grass-Fed Beef: Market Share of Grass-Fed Beef," South Dakota State University Extension, last modified June 28, 2021, https://extension.sdstate.edu/grass-fed-beef-market-share-grass-fed-beef.

238 Michaeleen Doucleff, "Between Pigs and Anchovies: Where Humans Rank on the Food Chain," *The Salt* (*blog*), NPR, December 8, 2013, https://www.npr.org/sections/thesalt/2013/12/08/249227181/between-pigs-and-anchovies-where-humans-rank-on-the-food-chain.

239 United Nations, "Plants, the 'Core Basis for Life on Earth,' Under Increasing Threat, Warns UN Food Agency," UN News, December 2, 2019, https://news.un.org/en/story/2019/12/1052591.

240 Doucleff, "Between Pigs and Anchovies."

241 Adam Dave, "Duration of Digestion: Meat Diet vs. Vegetarian Diet," LiveStrong.com, last modified November 9, 2018, https://www.

livestrong.com/article/441273-how-long-does-it-take-a-meat-diet-to-digest-compared-to-a-vegetarian-one.

242 Brian Pobiner, "Why Do We Eat Meat? Tracing the Evolutionary History," interview by Lynne Rossetto Kasper, *The Splendid Table*, March 6, 2013, https://www.splendidtable.org/story/2013/03/06/why-do-we-eat-meat-tracing-the-evolutionary-history.

243 Caitlin O'Kane, "Fair Oaks Farms Under Investigation After Undercover Video Exposes Animal Abuse," CBS News, June 7, 2019, https://www.cbsnews.com/news/after-undercover-video-exposes-animal-abuse-at-fair-oaks-farms-grocery-store-removes-products.

244 "Shredding Day-Old Chicks: How Australia's Egg Industry Works," Triple J Hack, July 20, 2016, https://www.abc.net.au/triplej/programs/hack/chickens/7645698.

245 People for the Ethical Treatment of Animals (PETA), "Aqua-farming," accessed March 11, 2022, https://www.peta.org/issues/animals-used-for-food/factory-farming/fish/aquafarming.

246 "Fish Slaughter," accessed April 11, 2022, https://dbonus869y26v.cloudfront.net/en/Fish_slaughter.

247 Farm Animal Welfare Council, "Report on the Welfare of Farmed Fish—Recommendations: Trout," July 28, 2012, https://web.archive.org/web/20120728115938/http://www.fawc.org.uk/reports/fish/fishr085.htm.

248 Peter Singer, "Fish: The Forgotten Victims on Our Plate," *The Guardian*, September 14, 2010, https://www.theguardian.com/commentisfree/cif-green/2010/sep/14/fish-forgotten-victims.

249 Heather Browning and Walter Veit, "Is Humane Slaughter Possible?" *Animals* 10, no. 5 (May 2020), https://doi.org/10.3390/ani10050799.

250 Elizabeth Burke, "Why Use Zebrafish to Study Human Diseases?" NIH Intramural Research Program, August 9, 2016, https://irp.nih.gov/blog/post/2016/08/why-use-zebrafish-to-study-human-diseases.

251 Matt Parker, "5 Ways Fish are Like You and Me," EarthSky, April 8, 2021, https://earthsky.org/earth/ways-fish-are-like-humans.

252 Animal Welfare Act, 7 U.S.C. ch. 54 § 2131–2132 (1966).

253 Packaged Facts, "Animal Welfare: Issues and Opportunities in the Meat, Poultry, and Eggs Markets in the U.S.," Report (April 10, 2017), https://www.packagedfacts.com/Animal-Welfare-Meat-10771767/?progid=89555.

254 Animal Crush Video Prohibition Act, 111–294 (2010), https://www.govinfo.gov/link/statute/124/3177.

255 Preventing Animal Cruelty and Torture (PACT) Act, 116–72 (2019), https://www.congress.gov/bill/116th-congress/house-bill/724/text.

256 "Number of Animals Slaughtered," Our World in Data, accessed April 11, 2022, https://ourworldindata.org/meat-production#number-of-animals-

slaughtered. See also "Factory Farming," A Well Fed World, accessed March 11, 2022, https://awellfedworld.org/factory-farms.

257 Animal Clock, "Questions and Answers," accessed May 17, 2022, https://animalclock.org.

258 "Reducing Suffering in Fisheries," FishCount.org, accessed March 11, 2022, http://fishcount.org.uk.

259 C. Victor Spain et al., "Are They Buying It? United States Consumers' Changing Attitudes toward More Humanely Raised Meat, Eggs, and Dairy," *Animals* 8, no. 8 (2018), https://dx.doi.org/10.3390%2Fani8080128.

260 "About Esther," Esther the Wonder Pig, accessed April 11, 2022, estherthewonderpig.com/about.

261 Robert E. Black et al., "Maternal and Child Undernutrition and Overweight in Low-Income and Middle-Income Countries," *The Lancet* 382, no. 9890 (August 2013), https://doi.org/10.1016/S0140-6736(13)60937-X.

262 Shiva, *Stolen Harvest*, 70–71.

263 Michael Klaper, interview with David Horton, All-Creatures.org, September 2010, https://www.all-creatures.org/articles/klaper.html.

264 Erin R. Hahn, Meghan Gillogly, and Bailey E. Bradford, "Children are Unsuspecting Meat Eaters: An Opportunity to Address Climate Change," *Journal of Environmental Psychology* 78 (December 2021): 6, https://doi.org/10.1016/j.jenvp.2021.101705.

265 Luke McGuire, Sally B. Palmer, and Nadira S. Faber, "The Development of Speciesism: Age-Related Differences in the Moral View of Animals," *Social Psychological and Personality Science* (April 2022): 1, https://doi.org/10.1177%2F19485506221086182.

266 McGuire, Palmer, and Faber, "The Development of Speciesism," 9.

267 McGuire, Palmer, and Faber, "The Development of Speciesism," 8.

268 Lori Marino and Christina M. Colvin, "Thinking Pigs: A Comparative Review of Cognition, Emotion, and Personality in *Sus domesticus*," *International Journal of Comparative Psychology* 28 (2015): 5, https://escholarship.org/uc/item/8sx4s79c.

269 James Gorman, "Dog Breeding in the Neolithic," *The New York Times*, June 25, 2020, https://www.nytimes.com/2020/06/25/science/arctic-sled-dogs-genetics.html.

270 Lori Marino, "Thinking Chickens: A Review of Cognition, Emotion, and Behavior in the Domestic Chicken," *Animal Cognition* 20 (2017), https://doi.org/10.1007/s1007-016-1064-4.

271 Lori Marino and Kristin Allen," The Psychology of Cows," *Animal Behavior and Cognition* 4, no. 4 (2017), https://dx.doi.org/10.26451/abc.04.04.06.2017.

272 Douglas Elliffe and Lindsay Matthews, "We Managed to Toilet Train

Cows (and They Learned Faster than a Toddler). It Could Help Combat Climate Change," *The Conversation*, September 13, 2021, https://the-conversation.com/we-managed-to-toilet-train-cows-and-they-learned-faster-than-a-toddler-it-could-help-combat-climate-change-167785.

273 Elliffe and Matthews, "We Managed to Toilet Train."

274 Jessica Pierce, "Do Animals Experience Grief?" *Smithsonian Magazine*, August 24, 2018, https://www.smithsonianmag.com/science-nature/do-animals-experience-grief-180970124.

275 Krista M. McLennan, "Social Bonds in Dairy Cattle: The Effect of Dynamic Group Systems on Welfare and Productivity," (doctoral thesis, University of Northampton, 2013), http://nectar.northampton.ac.uk/6466/1/McLennan_Krista_2013_Social_bonds_in_dairy_cattle_the_effect_of_dynamic_group_systems_on_welfare_and_productivity.pdf.

276 Charlotte Gaillard et al., "Social Housing Improves Dairy Calves' Performance in Two Cognitive Tests," *PLoS ONE* 9, no. 2 (Februrary 2014), https://doi.org/10.1371/journal.pone.0090205.

277 Jordana Cepelewicz, "Animals Count and Use Zero. How Far Does Their Number Sense Go?" *Quanta Magazine*, August 9, 2021, https://www.quantamagazine.org/animals-can-count-and-use-zero-how-far-does-their-number-sense-go-20210809.

278 Maximillian E. Kirschhock, Helen M. Ditz, and Andreas Nieder, "Behavioral and Neuronal Representation of Numerosity Zero in the Cow," *Journal of Neuroscience* 41, no. 22 (June 2021), https://doi.org/10.1523/JNEUROSCI.0090-21.2021.

279 Cepelewicz, "Animals Count and Use Zero."

280 Rosa Rugani et al., "Arithmetic in Newborn Chicks," *Proceedings of the Royal Society B* 276, no. 1666 (July 2009), https://doi.org/10.1098/rspb.2009.0044.

281 Simon Maybin, "Busting the Attention Span Myth," BBC World Service, March 10, 2017, https://www.bbc.com/news/health-38896790.

282 Rajvi Desai, "Why We Are Able to Empathize with Some, Never All," The Swaddle, July 7, 2019, https://theswaddle.com/selective-empathy-why-we-are-able-to-empathize-with-some-never-all.

283 Robert J. Lifton, "Cult Formation," *Cultic Studies Journal* 8, no. 1 (1991), https://psycnet.apa.org/record/1992-12427-001.

284 *The Game Changers*, directed by Louie Psihoyos (2018), https://gamechangersmovie.com.

285 Mark Z. Johnson, "Beef—Is it What's for Dinner?" *Beef Magazine*, November 22, 2021, https://www.beefmagazine.com/beef/beef-it-whats-dinner.

286 Sarah Miller, "Eating Meat is Inhumane, Bad for the Envi-

ronment, and Harmful to My Heath. I Still Can't Give It
Up," *Insider*, March 10, 2020, https://www.insider.com/
why-i-eat-meat-even-though-inhumane-bad-for-health-2020-3.

287   Evelyn Medawar et al., "The Effects of Plant-Based Diets on the Body
and the Brain: A Systematic Review," *Traditional Psychiatry* 9 (2019),
https://doi.org/10.1038/s41398-019-0552-0.

288   Ross, "Watch Out for Tell-tale Signs."

289   Simon Armon, "What's It Like Collecting Bull Cum for a Living?"
interview by Sam Nichols, *VICE*, June 29, 2017, https://www.vice.com/en/
article/qv4y8b/whats-it-like-collecting-bull-cum-for-a-living.

290   "Bovine Semen: Imports and Exports 2020", Trend Economy, accessed
April 12, 2020, https://trendeconomy.com/data/commodity_h2/051110.

291   Dave Rogers, "Strange Noises Turn Out to Be Cows Missing Their
Calves," *The Daily News*, October 23, 2013, https://www.newburyport-
news.com/news/local_news/strange-noises-turn-out-to-be-cows-miss-
ing-their-calves/article_d872e4da-b318-5e90-870e-51266f8eea7f.html.

292   University of Veterinary Medicine, Vienna," Early Separation of Cow and
Calf Has Long-Term Effects on Social Behavior," *ScienceDaily*, April 28,
2015, https://www.sciencedaily.com/releases/2015/04/150428081801.htm.

293   Mary Bates, "The Emotional Lives of Dairy Cows," *WIRED*, June 30,
2014, https://www.wired.com/2014/06/the-emotional-lives-of-dairy-cows.

294   See for example, "Calf Slaughter: The Killing of Baby Cows," Kinder
World, accessed April 12, 2022, https://www.kinderworld.org/videos/
dairy-industry/calf-slaughter.

295   Darolyn L. Stull et al., "A Review of the Causes, Prevention, and Welfare
of Nonambulatory Cattle," *Journal of the American Veterinary Medical
Association* 231, no. 2 (July 2007), https://doi.org/10.2460/javma.231.2.227.

296   Livestock and Poultry Environmental Learning Community Admin,
"Liquid Manure Storage Ponds, Pits, and Tanks," Livestock and Poultry
Environmental Learning Community, March 5, 2019, https://lpelc.org/
liquid-manure-storage-ponds-pits-and-tanks.

297   Ritchie and Roser, "Meat and Dairy Production." See also P. Sans and P.
Combris, "World Meat Consumption Patterns: An Overview of the Last
Fifty Years (1951–2011)," *Meat Science* 109 (November 2015), https://doi.
org/10.1016/j.meatsci.2015.05.012.

298   Laura Garnham and Hanne Lovlie, "Sophisticated Fowl: The Complex
Behaviour and Cognitive Skills of Chickens and Red Junglefowl," *Behavioral Sciences* 8, no. 13 (January 2018), https://doi.org/10.3390/bs8010013.

299   Brian L. Chen, Kathryn L. Haith, and Bradley A. Mullens, "Beak Condition Drives Abundance and Grooming-Mediated Competitive Asymme-
try in a Poultry Ectoparasite Community," *Parasitology* 138, no. 6 (March

2011), https://doi.org/10.1017/S0031182011000229.

300 *Dominion*, directed by Chris Delforce (2018), https://www.dominion-movement.com/watch.

301 Arun Rath, "Ringling Brothers Official Retires Circus Elephants," *All Things Considered* on NPR, May 2, 2016, https://www.npr.org/2016/05/02/476498700/ringling-brothers-officially-retires-circus-elephants.

302 James Zoltak, "Feld Entertainment Is Bringing Ringling Bros. and Barnum & Bailey Circus Back," Venues Now, October 21, 2021, https://venuesnow.com/feld-entertainment-is-bringing-the-circus-back.

303 "Montana Animal Trap Restrictions Initiative I-177 (2016)," BallotPedia, accessed March 15, 2022, https://ballotpedia.org/Montana_Animal_Trap_Restrictions_Initiative,_I-177_(2016).

304 Refuge from Cruel Trapping Act, H.R. 4716, 117th Congress (2021), https://www.congress.gov/bill/117th-congress/house-bill/4716.

305 Animal Welfare Institute, "Refuge from Cruel Trapping Act Would Protect Animals, Wild and Domestic," July 27, 2021, press release, https://awionline.org/press-releases/refuge-cruel-trapping-act-would-protect-animals-wild-and-domestic.

306 Stephanie Simon, "In a Beef Over Branding," *The Wall Street Journal*, May 24, 2011, https://www.wsj.com/articles/SB10001424052702303654804576341303811402170.

307 Kathleen Stachowski, "Cattle Branding: Tradition Without a Heart," *Encyclopedia Britannica*, accessed March 17, 2022, https://www.britannica.com/explore/savingearth/hot-iron-cattle-branding-tradition-without-a-heart.

308 Katy Severson, "This Magic Ratio of Fat to Carbs Makes the Perfect Comfort Food," *HuffPost*, February 4, 2019, https://www.huffpost.com/entry/ratio-fat-carbs-comfort-food_l_5c530c48e4b0ca92c6de1760.

309 "Umami Basics," Umami Information Center, accessed March 17, 2022, https://www.umamiinfo.com/what/whatisumami.

310 N. Chaudhari, A.M. Landin, and S.D. Roper, "A Metabotropic Glutamate Receptor Variant Functions as a Taste Receptor," *Nature Neuroscience* 3, no. 2 (February 2000), https://doi.org/10.1038/72053.

311 "A Taste for Fate May Have Made Us Human," *ScienceDaily*, February 5, 2019, https://www.sciencedaily.com/releases/2019/02/190205161420.htm.

312 FEMA Staff, "What Flavors Meat and Why We Crave It," Flavor and Extract Manufacturers Association, accessed March 17, 2022, https://www.femaflavor.org/what-flavors-meat-and-why-we-crave-it.

313 Gerhard Feiner, *Meat Products Handbook: Practice Science and Technology* (Cambridge: Woodhead Publishing, 2006), 70.

314 Marta Zaraska, "What Makes a Hamburger and Other Cooked Meat So
     Enticing to Humans?" *The Washington Post*, August 12, 2013, https://www.
     washingtonpost.com/national/health-science/what-makes-a-hamburger-
     and-other-cooked-meat-so-enticing-to-humans/2013/08/12/8f8e1d72-
     ff73-11e2-9711-3708310f6f4d_story.html.

315 Mary Jane Brown, "What Are Advanced Glycation End
     Products (AGEs)?" Healthline, last modified Octo-
     ber 22, 2019, https://www.healthline.com/nutrition/
     advanced-glycation-end-products#what-they-are.

316 Brown, "What Are Advanced Glycation."

317 Brown, "What Are Advanced Glycation."

318 ASPCA, "ASPCA Research Shows Americans."

319 Farhad Manjoo, "I Can't Stop Wondering What's Going on Inside My
     Cat's Head," *The New York Times* Opinion, August 27, 2021, https://www.
     nytimes.com/2021/08/27/opinion/cats-dogs-consciousness.html.

320 Philip Low, "The Cambridge Declaration on Consciousness," presented
     at the Francis Crick Memorial Conference on Consciousness in Human
     and non-Human Animals, Churchill Collee, University of Cambridge,
     July 7, 2012, https://fcmconference.org/img/CambridgeDeclarationOn-
     Consciousness.pdf.

321 "Octopuses, Crabs and Lobsters to be Recognised as Sentient
     Beings Under UK Law Following LSE Report Findings," Lon-
     don School of Economics, November 19, 2021, https://www.
     lse.ac.uk/News/Latest-news-from-LSE/2021/k-November-21/
     Octopuses-crabs-and-lobsters-welfare-protection.

322 Lori Marino, "Eating Someone," *Aeon*, May 8, 2019, https://aeon.co/
     essays/face-it-a-farmed-animal-is-someone-not-something.

323 "The Foundations of Empathy are Found in the Chicken,"
     myScience,org, March 9, 2011, https://www.myscience.org/news/2011/
     the_foundations_of_empathy_are_found_in_the_chicken-2011-bristol.

324 Marino, "Eating Someone."

325 Gillian Wong, "China Dog Rescue: Hundreds of Animals Rescued
     from Slaughter by Activist Road Blockade," HuffPost, April 20,
     2011, updated December 6, 2017, https://www.huffpost.com/entry/
     dog-rescue-china-animal-activists_n_851454.

326 Chelsea Todaro, "Florida Man Gets Year in Jail for Running Over
     9 Ducklings with Lawnmower," *The Atlanta Journal-Constitution*,
     July 25, 2015, https://www.ajc.com/news/national/florida-man-
     gets-year-jail-for-running-over-ducklings-with-lawnmower/
     H8McZRq7O9WzaoENduedeO.

327 Hank Rothgerber, "Efforts to Overcome Vegetarian-Induced Dissonance

Among Meat Eaters," *Appetite* 79 (August 2014), https://doi.org/10.1016/j.appet.2014.04.003.

328 Animals Australia team, "The Reality of Farming Animals: Painful Mutilations," Animals Australia, last modified November 3, 2020, https://animalsaustralia.org/latest-news/reality-farming-animals-painful-mutilations.

329 United States Department of Agriculture, "Death Loss in U.S. Cattle and Calves Due to Predator and Nonpredator Causes, 2015," USDA, December 2017, https://www.aphis.usda.gov/animal_health/nahms/general/downloads/cattle_calves_deathloss_2015.pdf.

330 USDA, "Death Loss in U.S. Cattle."

331 Carol J. Adams, "The Inherently Bad Logic of the 'Only One Bad Day' Claim," *Carol J. Adams* (blog), September 14, 2015, https://caroljadams.com/carol-adams-blog/the-inherently-bad-logic-of-the-only-one-bad-day-claim.

332 Derrick Z. Jackson, "Food Companies at the Table in Trump Administration's Dietary Guidelines Committee," *The Equation*, Union of Concerned Scientists, April 1, 2019, https://blog.ucsusa.org/derrick-jackson/trump-administrations-dietary-guidelines-committee.

333 U.S. Department of Health and Human Services and U.S. Department of Agriculture, 2015–2020 *Dietary Guidelines for Americans*, 8th Edition (December 2015), available at https://health.gov/our-work/nutrition-physical-activity/dietary-guidelines/previous-dietary-guidelines/2015.

334 Consolidated Appropriations Act, H.R. 2029, 114th Congress, §734 (2016).

335 US Department of Agriculture, "National School Lunch Program," Economic Research Service USDA, updated January 14, 2022, https://www.ers.usda.gov/topics/food-nutrition-assistance/child-nutrition-programs/national-school-lunch-program.

336 Zaraska, *Meathooked*, 8.

337 Health and Human Services and US Department of Agriculture, 2015–2020 *Dietary Guidelines for Americans* (December 2015): 40, https://health.gov/our-work/nutrition-physical-activity/dietary-guidelines/previous-dietary-guidelines/2015.

338 US Department of Agriculture, "Nutrient Intakes from Food and Beverages: Mean Amounts Consumed per Individual, by Gender and Age, in United States, 2015–2016," in the What We Eat in America Database, al Heatth and Nutrition Examination Survey, https://www.ars. arsuserfiles/80400530/pdf/1516/table_1_nin_gen_15.pdf. ant of Agriculture, *Scientific Report of the 2020 Dietary Guide-*

lines Advisory Committee (July 2020): 46, https://www.dietaryguidelines.gov/sites/default/files/2020-07/ScientificReport_of_the_2020DietaryGuidelinesAdvisoryCommittee_first-print.pdf.

340 Center for Disease Control and Prevention, "Men and Heart Disease," reviewed February 3, 2021, https://www.cdc.gov/heartdisease/men.htm.

341 Esselstyn et al., "A Way to Reverse CAD?"

342 Jeanine Bentley, "Trends in U.S. Per Capita Consumption of Dairy Products, 1970–2012," U.S. Department of Agriculture, Economic Research Service, June 2, 2014, https://www.ers.usda.gov/amber-waves/2014/june/trends-in-us-per-capita-consumption-of-dairy-products-1970-2012; M. Shahbandeh, "U.S. per Capita Consumption of Cheese 2000–2020," Statista, November 5, 2021, https://www.statista.com/statistics/183785/per-capita-consumption-of-cheese-in-the-us-since-2000.

343 HHS and USDA, *2015–2020 Dietary Guidelines for Americans*, 18.

344 Jeanine Bentley, "U.S. Trends in Food Availability and a Dietary Assessment of Loss-Adjusted Food Availability, 1970–2014," Economic Information Bulletin No. (EIB–166), U.S. Department of Agriculture, Economic Research Service, January 2017, https://www.ers.usda.gov/publications/pub-details/?pubid=82219.

345 Walter C. Willett and David S. Ludwig, "Milk and Health," *The New England Journal of Medicine* 382 (2020), https://doi.org/10.1056/NEJMra1903547.

346 Anthony Rivas, "Three Daily Servings: Could Reduced-Fat Milk Contribute to Obesity?" Medical Daily, July 3, 2013, https://www.medicaldaily.com/three-daily-servings-could-reduced-fat-milk-contribute-obesity-247392.

347 Willett and Ludwig, "Milk and Health."

348 "What Are Clinical Trials and Studies?" National Institute of Health: National Institute on Aging, reviewed April 9, 2020, https://www.nia.nih.gov/health/what-are-clinical-trials-and-studies.

349 Willett and Ludwig, "Milk and Health."

350 University of Technology, Sydney, "The Link Between Meat and Social Status," Phys.org, September 7, 2018, https://phys.org/news/2018-09-link-meat-social-status.html.

351 National Chicken Council, "Per Capita Consumption of Poultry and Livestock, 1965 to Forecast 2022, in Pounds," last modified December 2021, https://www.nationalchickencouncil.org/about-the-industry/statistics/per-capita-consumption-of-poultry-and-livestock-1965-to-estimated-2012-in-pounds.

352 Ned Stafford, "History: The Changing Notion of Food," *Nature* 468 (December 2010), https://doi.org/10.1038/468S16a.

353 Simon, *Meatonomics*, 28. See also Wilbur Olin Atwater, "Foods: Nutritive Value and Cost," U.S. Department of Agriculture Farmer's Bulletin No. 23 (pamphlet, 1894): 17–18, accessed April 14, 2022, University of North Texas Digital Library, https://digital.library.unt.edu/ark:/67531/metadc85506.

354 T. Colin Campbell, "The Protein Juggernaut Has Deep Roots," T. Colin Campbell Center for Nutrition Studies, last modified January 3, 2019, https://nutritionstudies.org/protein-juggernaut-deep-roots.

355 Anders Gaarn du Jardin Nielsen and Neil H. Metcalfe, "Mikkel Hindhede (1862–1945): A Pioneering Nutritionist," *Journal of Biomedical Biography* 26, no. 3 (2018), https://doi.org/10.1177/0967772015623412.

356 du Jardin Nielsen and Metcalfe, "Mikkel Hindhede."

357 Michael R. MacArthur, "Total Protein, Not Amino Acid Composition, Differs in Plant-Based Versus Omnivorous Dietary Patterns and Determines Metabolic Health Effects in Mice," *Cell Metabolism* 33, no. 9 (July 2021): 1808, https://doi.org/10.1016/j.cmet.2021.06.011.

358 Mical Dorfner, "Are You Getting Too Much Proetin," Mayo Clinic News Network, February 23, 2017, https://newsnetwork.mayoclinic.org/discussion/are-you-getting-too-much-protein.

359 T. Brock Symons et al., "A Moderate Serving of High-Quality Protein Maximally Stimulates Skeletal Muscle Protein Synthesis in Young and Elderly Subjects," *Journal of the Academy of Nutrition and Dietetics* 109, no. 9 (September 2009): 1582, https://doi.org/10.1016/j.jada.2009.06.369.

360 Kris Gunnar, "Protein Intake—How Much Protean Should You Eat per Day?" Healthline, October 1, 2020, https://www.healthline.com/nutrition/how-much-protein-per-day.

361 Vesanto Melina, Winston Craig, and Susan Levin, "Position of the Academy of Nutrition and Dietetics: Vegetarian Diets," *Journal of the Academy of Nutrition and Dietetics* 166, no. 12 (December 2016): 1971, https://doi.org/10.1016/j.jand.2016.09.025.

362 Christopher D. Gardner et al., "Maximizing the Intersection of Human Health and the Health of the Environment with Regard to the Amount and Type of Protein Produced and Consumed in the United States," *Nutrition Reviews* 77, no. 4 (April 2019): 197–215, https://doi.org/10.1093/nutrit/nuy073.

363 Meng T. Lim et al., "Animal Protein versus Plant Protein in Supporting Lean Mass and Muscle Strength: A Systematic Review and Meta-Analysis ᶜ Randomized Controlled Trials," *Nutrients* 13, no. 2 (2021): 661, https://ᵪ/10.3390/nu13020661.

    ·t al., "Maximizing the Intersection," 198.

    ·grams—An Overview," The National Agricultural Law

Center, accessed March 28, 2022, https://nationalaglawcenter.org/overview/checkoff.

366 Zaraska, *Meathooked*, 91.

367 Simon, *Meatonomics*, 9.

368 "Beef Checkoff Showcases Great Value to Beef Industry in ROI Analysis," Beef Board, October 21, 2019, https://www.beefboard.org/2019/10/21/beef-checkoff-showcases-great-value-to-beef-industry-in-roi-analysis; Kevin Schulz, "Report: Pork Checkoff Pays Off for Pork Producers," National Hog Farmer, January 8, 2018, https://www.nationalhogfarmer.com/business/report-pork-checkoff-pays-pork-producers; Gary W. Williams and Dan Hanselka, "Return on Investment in the American Lamb Checkoff Program," Report to the American Lamb Board, February 2019, https://static1.squarespace.com/static/5df2bf866e42151b88818a9e/t/5e98d0293d79a11584a5e2bd/1587073066541.

369 "Hog Farmers Criticize Veneman for not Terminating Checkoff," *High Plains Journal*, January 1, 2001, https://www.hpj.com/archives/hog-farmers-criticize-veneman-for-not-terminating-checkoff/article_3c8c6451-5341-55b7-9c70-974e3d5435d9.html.

370 "Checkoff Programs—An Overview."

371 Simon, *Meatonomics*, 6.

372 North American Meat Institute, "2019 Economic Impact of the Meat and Poultry Industry," accessed April 12, 2022, https://www.meatinstitute.org/index.php?ht=d/sp/i/156726/pid/156726.

373 Zaraska, *Meathooked*, 91.

374 Chase Purdy, "The Secretive US Funding Behind the Got Milk? Ads Finally Gets Scrutiny," *Quartz*, March 29, 2017, https://qz.com/944630/the-us-government-program-behind-the-got-milk-campaign-attacked-vegan-mayo-and-is-now-under-congressional-scrutiny.

375 Purdy, "Secretive US Funding."

376 Purdy, "Secretive US Funding."

377 "People Who Eat Meat Experience Lower Levels of Depression and Anxiety Compared to Vegans," *Sapien Journal*, November 28, 2021, https://sapienjournal.org/people-who-eat-meat-experience-lower-levels-of-depression-and-anxiety-compared-to-vegans.

378 Urska Dobersek, "Meat and Mental Health: A Meta-Analysis of Meat Consumption, Depression, and Anxiety," *Critical Reviews in Food Science and Nutrition* (2021), https://doi.org/10.1080/10408398.2021.1974336.

379 Agricultural Fairness Alliance, "Beef," April 19, 2021, https://www.agriculturefairnessalliance.org/news/beef.

380 Simone Spearman, "Eating More Veggies Can Help Save Energy," *SFGATE* opinion, June 29, 2001, https://www.sfgate.com/opinion/open-

forum/article/Eating-more-veggies-can-help-save-energy-2904772.php.

381 Kentucky-American Water Company, "Rates – Charges – Rules – Regulations for Furnishing Water Service," effective June 28, 2019, https://www.amwater.com/kyaw/resources/PDF/KAW_water%20rates%20only.pdf.

382 Raj Patel, *The Value of Nothing: How to Reshape Market Society and Redefine Democracy* (New York: Picador, 2010).

383 Tyson Foods, "Beyond Meat and Tyson Foods Announce Investment Agreement," Press Release, October 10, 2016, https://www.tysonfoods.com/news/news-releases/2016/10/beyond-meat-and-tyson-foods-announce-investment-agreement.

384 "The Dairy that Gave Up Dairy: Our Story," Elmhurst, accessed March 18, 2022, https://elmhurst1925.com/pages/our-story.

385 "NHE Fact Sheet," CMS.gov, last modified December 15, 2021, https://www.cms.gov/Research-Statistics-Data-and-Systems/Statistics-Trends-and-Reports/NationalHealthExpendData/NHE-Fact-Sheet.

386 Majid Ezzati et al., "Rethinking the 'Diseases of Affluence' Paradigm: Global Patterns of Nutritional Risks in Relation to Economic Development," *PLoS Medicine* 2, no. 5 (May 2005), https://doi.org/10.1371/journal.pmed.0020133.

387 Zaraska, *Meathooked*, 95.

388 Josefina Salomon, "Could the Economic Disaster Turn Meat-Loving Argentines Vegan?" OZY, January 1, 2020, https://www.ozy.com/the-new-and-the-next/could-the-economic-crisis-turning-meat-loving-argentinians-vegan/252699.

389 Vesanto Melina, Winston Craig, and Susan Levin, "Position of the Academy of Nutrition and Dietetics: Vegetarian Diets," *Journal of the Academy of Nutrition and Dietetics* 116, no. 12 (December 2016): P1970, https://doi.org/10.1016/j.jand.2016.09.025.

390 Michael Greger, "Omnivore v. Vegan Nutrient Deficiencies," Nutrition Facts, video, 1:25, https://nutritionfacts.org/video/omnivore-vs-vegan-nutrient-deficiencies-2.

391 Michael Klaper et al., "Getting Started on a Whole Food Plant-Based Diet," Plant Based TeleHealth, 19:26, https://plantbasedtelehealth.com/2020/09/10/getting-started-on-a-whole-food-plant-based-diet.

ds High in Calcium for Vegans," Nourish by WebMD, ` March 18, 2022, https://www.webmd.com/diet/ `n-calcium-for-vegans#1.

`rganization, "Assessment of Iodine Deficiency Dis- `ing Their Elimination: A Guide for Programme

Managers," Third Edition (2007): 1, https://apps.who.int/iris/bitstream/handle/10665/43781/9789241595827_eng.pdf.

394 Rashmi Mullur, Yan-Yun Liu, and Gregory A. Brent, "Thyroid Hormone Regulation of Metabolism, *Physiological Reviews* 94, no. 2 (April 2014), https://doi.org/10.1152/physrev.00030.2013.

395 Bruce Neal, "Dietary Salt Is a Public Health Hazard That Requires Vigorous Attack," *Canadian Journal of Cardiology* 30, no. 5 (February 2014), https://doi.org/10.1016/j.cjca.2014.02.005; Stanley Shaldon and Joerg Vienken, "Beyond the Current Paradigm: Recent Advances in the Understanding of Sodium Handline—Guest Editors: Stanley Shaldon and Joerg Vienken: Salt, the Neglected Silent Killer," *Seminars in Dialysis* 22, no. 3 (June 2009), https://doi.org/10.1111/j.1525-139X.2009.00606.x.

396 Julieanna Hever, "Plant-Based Diets: A Physician's Guide," *The Permanente Journal* 20, no. 3 (Summer 2016), https://doi.org/10.7812/TPP/15-082.

397 NIH Office of Dietary Supplements, "Iodine: Fact Sheet for Health Professionals," NIH, updated March 1, 2022, https://ods.od.nih.gov/factsheets/Iodine-HealthProfessional.

398 Lindsay H. Allen, "How Common is Vitamin B-12 Deficiency," *The American Journal of Clinical Nutrition* 89, no. 2 (February 2009): 693S, https://doi.org/10.3945/ajcn.2008.26947A.

399 Parabel USA Inc., "Parabel Announces Natural Plant Source of Vitamin B12 in Water Lentils and LENTEIN Plant Protein," press release, PR Newswire, November 18, 2019, https://www.prnewswire.com/news-releases/parabel-announces-natural-plant-source-of-vitamin-b12-in-water-lentils-and-lentein-plant-protein-300960037.html.

400 Lawton Stewart, "Mineral Supplements for Beef Cattle, Bulletin 895," University of Georgia Extension, reviewed March 31, 2017, https://extension.uga.edu/publications/detail.html?number=B895.

401 "Vitamin B12 Analogues," Vitamin B12 & Health, accessed March 18, 2022, https://www.b12-vitamin.com/analogues.

402 Marilee Nelson, "Mindful Eating: How Chewing Improves Digestion," *Branch Basics* (blog), March 3, 2018, https://branchbasics.com/blogs/food/mindful-eating-how-chewing-improves-digestion.

403 Raj Kishor Gupta, Shivraj Singh Gangoliya, and Nand Kumar Singh, "Reduction in Phytic Acid and Enhancement of Bioavailable Micronutrients in Food Grains," *Journal of Food Science and Technology* 52 (2015), https://doi.org/10.1007/s13197-013-0978-y.

404 Weston Petroski, "Is There Such a Thing as 'Anti-Nutrient'? A Narrative Review of Perceived Problematic Plant Compounds," *Nutrients* 12, no. 10 (2020): 15, https://doi.org/10.3390/nu12102929.

405 Petroski, "Is there Such a Thing as 'Anti-Nutrient'?"

406 "What Are the Health Benefits (and Risks) of Eating Raw Sprouts?" Cleveland Clinic Health Essentials, March 3, 2021, https://health.cleve-landclinic.org/what-are-the-health-benefits-and-risks-of-eating-sprouts.

407 WebMD Editorial Contributors, "Sprouts: Are They Good for You?" Nourish by Web MD, reviewed September 29, 2020, https://www.webmd.com/diet/sprouts-good-for-you.

408 "What Are the Health Benefits?"

409 Verena Tan, "How to Increase the Absorption of Iron From Foods," Healthline, June 3, 2017, https://www.healthline.com/nutrition/increase-iron-absorption.

410 "Omega-3 Fates—Good for Your Heart," Medline Plus, reviewed May 26, 2020, https://medlineplus.gov/ency/patientinstructions/000767.htm.

411 National Institutes of Health Office of Dietary Supplements, "Omega-3 Fatty Acids," last modified August 4, 2021, https://ods.od.nih.gov/factsheets/Omega3FattyAcids-HealthProfessional.

412 NIH, "Omega-3 Fatty Acids."

413 Kat Gál, "What are the Best Sources of Omega-3?" *Medical News Today*, January 20, 2020, https://www.medicalnewstoday.com/articles/323144#_noHeaderPrefixedContent.

414 Artemis P. Simopoulos, "An Increase in the Omega-6/Omega-3 Fatty Acid Ratio Increases the Risk for Obesity," *Nutrients* 8, no. 3 (2016), https://doi.org/10.3390/nu8030128.

415 Dorna Davani-Davari et al., "Prebiotics: Definition, Types, Sources, Mechanisms, and Clinical Applications," *Foods* 8, no. 3 (March 2019), https://doi.org/10.3390/foods8030092.

416 Bethan Cadman, "What Prebiotic Foods Should People Eat?" *Medical News Today*, October 1, 2018, https://www.medicalnewstoday.com/articles/323214.

417 Bulsiewicz, *Fiber Fueled*.

418 Michael Klaper, "Meat Can Be Addictive According to Leading Doctor," interview by Klaus Mitchell, Plant Based News, February 11, 2019, https://plantbasednews.org/lifestyle/meat-addictive-leading-doctor.

419 "Read Meat, TMAO, and Your Heart," Harvard Health Publishing, September 1, 2019, https://www.health.harvard.edu/staying-healthy/red-meat-tmao-and-your-heart.

NOAA National Centers for Environmental Information, "State of Climate: National Climate Report for Annual 2021," January 2022, www.ncdc.noaa.gov/sotc/national/202113.

PCA Research Shows Americans."

ained

B/117